Lab Manual to Accompany Modern Livestock and Poultry Production

10th Edition

Frank B. Flanders
James R. Gillespie
Libby Hrlic Templeton

Cengage

Australia • Brazil • Canada • Mexico • Singapore • United Kingdom • United States

Cengage

***Lab Manual to Accompany Modern Livestock and Poultry Production*, 10th Edition**
Frank B. Flanders, James R. Gillespie and Libby Hrlic Templeton

SVP, Product Management: Cheryl Costantini

VP, Product Management: Audrey Turjanica

Portfolio Product Director: Jason Fremder

Associate Portfolio Product Manager: Nick Perry

Product Assistant: Eduarda David

Sr. Learning Designer: Jennifer Starr

Content Project Manager: Spandana Bakthavachala

Digital Project Manager: Scott Diggins

Development Editor: Janet Witalec, Katherine Pidde

Content Manager: Jalen Douglas

Sr. Product Marketing Manager: Caroline Cox

Director, Product Marketing: Adam Bolingbroke

Content Acquisition Analyst: Erin McCullough

Production Service: MPS Limited

Designer: Felicia Bennett

Cover Image Source: Dmitry Kalinovsky/Shutterstock.com; PeopleImages.com – Yuri A/Shutterstock.com; Mikhail Gnatkovskiy/Shutterstock.com; Eugene_Photo/Shutterstock.com; iStock.com/Arlee Piriyasanguanpon

Library of Congress Control Number: 2024942915

ISBN: 978-0-357-92446-4

Cengage
5191 Natorp Boulevard
Mason, OH 45040
USA

Cengage is a leading provider of customized learning solutions.
Our employees reside in nearly 40 different countries and serve digital learners in 165 countries around the world. Find your local representative at **www.cengage.com**.

To learn more about Cengage platforms and services, register or access your online learning solution, or purchase materials for your course, visit **www.cengage.com**.

Printed at CLDPC, USA, 8-26

Contents

Preface

The *Lab Manual* to accompany *Modern Livestock and Poultry Production*, Tenth Edition, is designed for agricultural education students who require competency in all phases and types of livestock production. The content contained in the *Lab Manual* is based on the most up-to-date information available and is applicable to all areas of the United States. This *Lab Manual*, complete with exercises and activities, accompanies the critical content areas covered throughout the book. The *Lab Manual* reinforces text content through the practical application of real-world examples. It is recommended that students complete corresponding lab components to further develop understanding of essential skills and concepts discussed in the text.

About the Author

Libby Hrlic Templeton has had an impressive career in agriculture education, serving the profession for over 20 years. Her dedication and accomplishments in agriculture, education, and leadership highlight her commitment to promoting agricultural education and providing opportunities for young people interested in agriculture. She taught agricultural education at Batesburg-Leesville High School for 16 years and is currently serving as the animal science teacher at the South Carolina Governor's School for Agriculture. She has served as President of the SC Agriculture Teachers Association and has coordinated the South Carolina Miss FFA Pageant for several years. Mrs. Templeton has had numerous FFA and 4-H state winning teams, two South Carolina State FFA officers, and a Three Star National Gold Chapter. Mrs. Templeton is a National FFA Ambassador and has served on leadership teams on many state and national organizations. Her extensive experience in raising and showing cattle, along with her success in leading FFA and 4-H state-winning leadership and livestock teams, speaks to her expertise and passion for agricultural education. Her academic achievements include a Bachelor of Science degree in Animal Science and a Master's Degree in Agriculture Education from Clemson University.

Section One

The Livestock Industry

Feel good studio/Shutterstock.com

Chapter 1

Overview and Development of the Livestock Industry

Introduction

The domestication of animals played a vital role in the development of civilization. Domesticated animals provided a dependable source of food and clothing. Most of the ancestors of present-day farm animals were first tamed in the Neolithic (New Stone) Age. Early explorers brought the various species of farm animals to the United States. These animals spread across the United States as early explorers and colonists moved westward. Certain states became leaders in livestock production due to factors such as climate, transportation routes, and access to natural resources.

Animals have many useful functions. They convert grain feeds into other food products, are a source of materials for clothing, and provide power and recreation. Animals concentrate bulky feeds, making them easier to market. Many by-products of animals are also important to society. Consumer concerns affect the trends in the livestock industries. Understanding the market inventory trends as well as consumer preferences is important for livestock business owners.

The most important consumer concern is the safety of our food products. The United States has some of the highest food safety regulations in the world. The U.S. Department of Agriculture (USDA) and the Food and Drug Administration (FDA) are responsible for creating and enforcing food safety laws. Several environmental factors can cause conditions that allow bacteria to grow on the food. According to the Centers for Disease Control and Prevention (CDC), the four major bacterial pathogens that contaminate meat and poultry products are *Salmonella, Campylobacter, E. coli* O157:H7, and *Listeria monocytogenes.*

The following lab exercises will focus on developing an understanding of the livestock industry, the history of animal domestication, analyzing market trends, and maintaining a safe supply of livestock products.

Name______________________________ Date______________

Exercise 1-1 Domestication Timeline Activity

For each of the animals commonly domesticated in the United States, fill in the blank squares on the timeline chart with proper information about the history of domestication for each species. You will need to include information on (1) classification, (2) the place and period of time each animal was first domesticated by humans, (3) major dates in the history of domestication in the United States, and (4) the key functions and uses of the animal by humans.

Domestication Timeline

Common Name	Family Genus Species	Time and Place of Origin	Arrival in the United States/Other Historical Dates	Function of Animal in Agriculture
Cattle				
Swine				
Sheep				
Goats				
Horses				
Poultry				

Name______________________________ Date_______________

Exercise 1-2 Class Activity: Leading States

Use Table 1-3 from the textbook to compare livestock production among states within the United States. The class will need to be divided into groups and each group assigned a category of livestock production. Each group should research the top three producing states from their category. They will need to give a presentation that explains why these states are suited to produce this livestock category. Using Internet resources, research the agricultural production statistics for the state, as well as other characteristics that may affect the ability to produce livestock products.

The following explains the criteria for the presentation:

Format: Each group's presentation will be given to the class using oral and visual content. The visual component may be developed using computers, apps, various software, or online resources.

Content: Each state should be described, highlighting characteristics that make it a good location to produce the livestock category. Examples of characteristics may be:

- geological attributes
- natural resources
- historical factors
- consumer demands
- transportation routes
- climatic factors
- population size and characteristics

Name____ Date__________________

Exercise 1-3 Trend Analysis Activity

The growth of livestock markets and the decrease in the number of farms have encouraged nonagricultural investors to seek out agricultural economic opportunities. Investors without agricultural experience rely on consultants for direction and instruction on how to invest their money.

For this exercise, you will take on the role of an agricultural business consultant. For Part 1, your task will be to analyze data provided throughout Chapter 1 in the textbook as well as other data found during online research. Using the analyzed data, in Part 2, you will prepare a proposal for your investor that highlights the trends in each livestock market and suggests which livestock category to invest in.

Using these data, create a line graph that shows the projected growth and decline of livestock populations from 1995 through the present day and beyond, depending on data availability (a graph template has been provided for you). Label the lines so it is easy to determine which animal each line represents.

Part 1: Complete the graphs using the appropriate data.

Graph # 1:

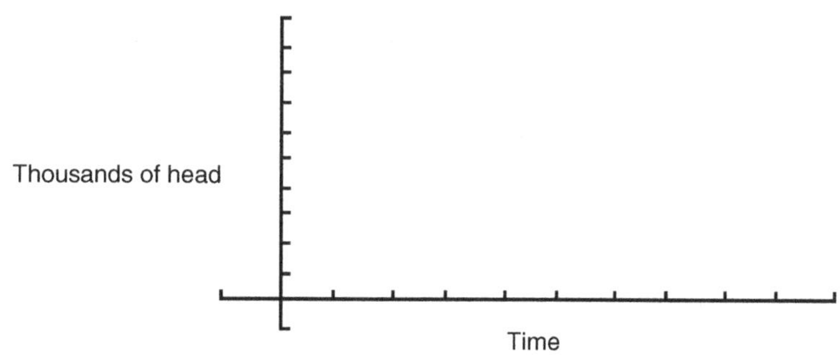

Graph # 2:

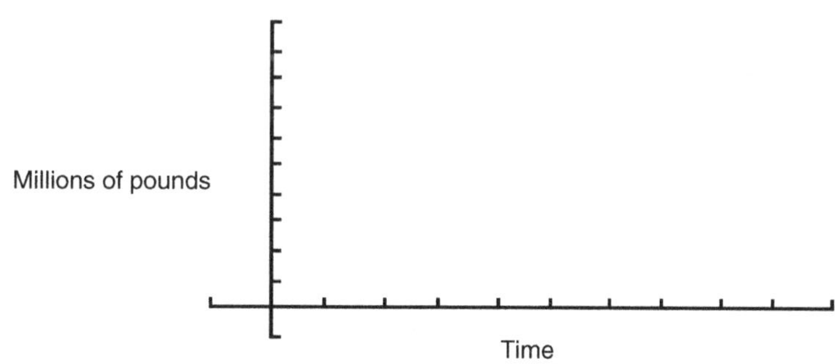

Part 2: For each of the livestock categories, describe the current market trends and projections for the future.

Beef Cattle Trends:

Current inventory numbers ____

Projected numbers ____

Market factors ____

Name____ Date____________________

Dairy Cattle Trends:

Current inventory numbers ____

Projected numbers ____

Market factors ____

Hog Trends:

Current inventory numbers ____

Projected numbers ____

Market factors ____

Chicken Trends:

Current inventory numbers ____

Projected numbers ____

Market factors ____

Turkey Trends:

Current inventory numbers ____

Projected numbers ____

Market factors ____

What is your final suggested livestock market?

____

____

____

____

Give your reasons for the choice.

____

____

____

____

Name______________________________________ Date__________________

Exercise 1-4 Food Safety

Foodborne illness is a serious concern facing all animal product markets. In recent years, the United States has improved its standards for sanitation and prevention of foodborne bacteria. All USDA-certified meat-processing plants in the United States must adhere to federal sanitation regulations and must operate with the presence of a USDA inspector. The system known as FAT TOM (an acronym for food, acidity, time, temperature, oxygen, and moisture) is used as a guide to understand environmental conditions that allow optimum growth of bacteria so that these conditions can be avoided. More information on FAT TOM can be found in the textbook.

Since 2000, all large meat-processing facilities have been required to comply with Hazard Analysis and Critical Control Point (HACCP) standards. There are seven principles that slaughterhouses in the United States must adhere to. The seven HACCP principles can be found in the textbook.

Food Safety Scenario

In this scenario, you, the student, will take the role of a USDA slaughterhouse inspector. Your job is to use the data that have been collected during a recent hazard analysis of a local meat-processing facility to apply the HACCP principles. Use the FAT TOM guidelines for reference as you conduct a review of the slaughter plant. If any violations are found, you must give recommendations for how the plant can become compliant as well as a timeline for making the changes.

The plant that is being reviewed has had some compliance issues in the past, so frequent inspections and strict enforcement of HACCP standards are necessary. During the most recent hazard analysis, the following environmental conditions were recorded:

> The hazard analysis was conducted on June 27; the temperature was 86 degrees. In the processing facilities, the temperature inside of the plant's walk-in cooler is normally 36 degrees. This temperature is maintained to preserve the carcasses as they hang prior to processing and packaging. On that morning, the compressor of the cooler malfunctioned, and it took 10 hours for it to be repaired. The cooler maintained a temperature below 41 degrees for 5 hours, after which the temperature entered the "danger zone." While this repair was being conducted, a puddle of stagnant water with a "wa" of 0.56 was found in the ceiling of the cooler; it had been leaking onto the carcasses as they hung. It was also found that a normal practice around the plant was to leave the cooler door open for hours at a time, letting outside air into the cooler. Another finding of concern was the time it would take the employees from the moment a beef carcass came out of the cooler until it was completely packaged in airtight packaging. That time was at least four hours. Another concern was the presence of stagnant water in several places throughout the facility. After samples were tested, the presence of salmonella and *E. coli* were found to be above the critical limits.

Food Safety Analysis

As the USDA inspector, create a report that identifies the violations of the HACCP guidelines as well as the steps and controls to ensure that the slaughterhouse can become compliant with the rules and avoid being shut down. Complete a

Name____ Date_____________________

HACCP analysis for the local slaughterhouse, using the seven principles discussed in the textbook (for example, schedule periodic testing, establish quarterly reviews).

1. Identify the findings of a hazard analysis.
2. Identify critical control points.
3. Establish critical limits for Critical Control Points (CCPs).
4. Establish requirements for monitoring the critical controls.
5. Identify the corrective actions to be taken.
6. Establish record-keeping procedures for CCPs.
7. Establish procedures to verify the system is working.

Name______________________________________ Date____________________

Chapter 1 Matching Activity

Directions: Match the term to the corresponding definition or description.

Term	Definition
____ 1. domesticate	**a.** extracted from the adrenal glands of animals and used for the treatment of hay fever and asthma
____ 2. *Bos taurus*	**b.** East Indian pig
____ 3. *Bos indicus*	**c.** animals, such as swine and poultry, that are fed large amounts of grain
____ 4. selection	**d.** to identify and breed animals with traits that are desirable by the breeder
____ 5. crossbreeding	**e.** animal used for pulling loads
____ 6. *Sus scrofa*	**f.** to adapt the behavior of an animal to fit the needs of people
____ 7. *Sus vittatus*	**g.** extracted from the adrenal glands of animals and used for the treatment of rheumatoid arthritis
____ 8. Mouflons	**h.** animals, such as cattle, sheep, and goats, that have a stomach that is divided into many compartments
____ 9. *Eohippus*	**i.** a tiny four-toed ancestor of the horse
____ 10. draft animal	**j.** wild big-horned sheep of Asia
____ 11. *Gallus gallus*	**k.** wild mallard duck; ancestor of all domestic breeds of ducks
____ 12. *Anas boschas*	**l.** extracted from the pancreas of animals and used in the treatment of diabetes
____ 13. non-ruminant	**m.** modern cattle descendant; humped cattle found in tropical countries
____ 14. ruminant	**n.** wild jungle fowl of India; ancestor of most tame chickens
____ 15. insulin	**o.** domestic cattle descending from the Aurochs or the Celtic Shorthorn
____ 16. cortisone	**p.** European wild boar
____ 17. thrombin	**q.** extracted from the blood of animals and used as a coagulant during surgery to help blood clot
____ 18. heparin	**r.** extracted from the stomachs of cattle and used to make cheese
____ 19. epinephrine	**s.** extracted from the lungs of animals and used during surgery to prevent blood clots and heart attacks
____ 20. rennet	**t.** mating of animals of different breeds

Name______________________________ Date______________

Chapter 1 Lab Questions

1. When evaluating hogs, beef, dairy, chickens, and turkeys, which market will have the most growth potential? Which market is second? Third?

2. Do animal rights and animal welfare have the same philosophy about animal use? What is the difference between the concerns of animal welfare and animal rights activists?

3. What are some factors that have led to the loss of genetic diversity in the livestock industry?

4. What are some factors that determine if food is safe to eat and free of harmful bacteria?

5. Discuss new trends that affect animal production.

Feel good studio/Shutterstock.com

Chapter 2

Career Opportunities in Animal Science

Introduction

Careers relating to agriculture are more relevant than ever—more than 300 different careers are available to persons with an interest in agriculture. Many of the career paths in animal science are not directly related to raising animals; for example, there are a variety of careers available from genetic engineering to selling grain to farmers. In the past 25 years, the number of jobs in production agriculture has decreased along with the number of full-time farms in the United States. The trend shows that since 1936, the number of farms has decreased, and the size of farms has increased; this trend is expected to continue. However, there has also been an increase in the number of small, part-time farms because of increased consumer demand for locally grown and organic products.

As the world's population grows rapidly, the demand for food is increasing as well. This demand is expanding the job market in careers that involve the processing, packaging, and marketing of food products. Many job opportunities in the animal industry are available in the fields of agribusiness, communications, science, government, education, sales, technology, engineering, and others. Previous experience working with animals is beneficial in any career path in the animal industry.

An increasing number of jobs that are available in animal science require a minimum education level of a bachelor's degree. In specialized areas of animal science, further education such as a master's degree or a PhD may be required. The number of job openings for college graduates with experience in agriculture over the next several years is expected to increase, while the number of qualified job applicants is not expected to keep up with the demand.

The objectives of the following exercises are to develop an understanding of the trends in the job market in the field of agriculture and to plan for the expected changes in employment opportunities in the future. You will conduct an employment opportunities assessment, research a career in animal agriculture, and prepare for an interview and job.

Name____________________ Date____________

Exercise 2-1 Employment Opportunities Assessment

Directions: For each career area given, list three potential job opportunities and describe one of the occupations. Use Tables 2-2 and 2-4 in the textbook for reference.

1. Career Area: **Scientists, Engineers, and Related Specialists**
   - Three examples of job opportunities:
   - Job description for one career:
2. Career Area: **Managers and Financial Specialists**
   - Three examples of job opportunities:
   - Job description for one career:
3. Career Area: **Marketing, Merchandising, and Sales Representatives**
   - Three examples of job opportunities:
   - Job description for one career:
4. Career Area: **Education, Communication, and Information Specialists**
   - Three examples of job opportunities:
   - Job description for one career:

Name______________________________ Date______________

5. Career Area: **Beef Occupations**
    - Three examples of job opportunities:
    - Job description for each career:

6. Career Area: **Dairy Occupations**
    - Three examples of job opportunities:
    - Job description for each career:

7. Career Area: **Swine Occupations**
    - Three examples of job opportunities:
    - Job description for each career:

8. Career Area: **Sheep and Goats Occupations**
    - Three examples of job opportunities:
    - Job description for each career:

9. Career Area: **Equine Occupations**
    - Three examples of job opportunities:

Name______________________________ Date______________

- Job description for each career:

____

____

10. Career Area: **Poultry Occupations**
    - Three examples of job opportunities:

    ____

    ____

    - Job description for each career:

    ____

    ____

Name____________________________________ Date____________________

Exercise 2-2 Career Exploration Activity

After researching the careers available in the field of animal agriculture, choose one career to research in more detail. Use the textbook and online resources to find information that describes the career you have chosen.

Task: Using your research, answer the following questions.

1. Name of the career/job you researched.

2. Area of animal science that the career fits into (livestock production, financial institutions, etc.).

3. Why did you select this career?

4. How/why is this career important in the production, marketing, or sales of animal products?

5. What skills and abilities are needed to be successful in this career (collaboration skills, communication skills, technical skills, etc.)?

6. What specialized skills are required?

7. What education is needed in high school to pursue this career?

8. What education is needed beyond high school?

Name____ Date____________________

9. What are the working conditions and situations that are present in this career?

____

____

10. What is the salary for someone starting versus someone with experience? Is it based on education level?

____

____

11. What are the advantages of this career?

____

____

12. What are the disadvantages of this career?

____

____

13. How can one find a job in this career?

____

____

14. Give three examples of current career opportunities in this field. Where are they located? What are the requirements of applicants?

____

____

15. Are you interested in pursuing this career? Why or why not?

____

____

Optional Exercise Extension Activity

Using the answers from questions 1–15, create a computer-based visual presentation on your chosen career. The presentation should be a minimum of eight slides, with a minimum of five visual aids. Use the answers to the questions for the content of the slides. Deliver a presentation to the class or a group of individuals who could be interested in this career field. The instructor will evaluate your presentation for accuracy and detail.

Name________________________________ Date__________________

Exercise 2-3 Job Interview Preparation

This exercise is modeled after the FFA Job Interviews Career Development Event (CDE). Your textbook, the internet, and the FFA CDE handbook are all appropriate resources for your research.

Tasks

As an individual, you need to prepare materials prior to applying for the career you have chosen. For this assignment, you will prepare a resume, a cover letter, a list of references, and a list of interview questions. You may also be responsible for a face-to-face mock interview with your teacher.

Note: For Parts 1, 2, and 3, use the same career that you described in Exercise 2-2.

Part 1: Resume

Your resume should consist of the following:

- Name and address
- Career objectives
- Education
- Skills
- Experience
- Achievements and projects
- Volunteer experience or community service
- Extracurricular activities
- References
- Any other important personal information

(An example resume is included on the next page for reference.)

Part 2: Cover Letter

Your cover letter should be addressed to a mock employer in your chosen career field. It should be typed, three paragraphs long, with your name and address in the heading. The letter should reference your skills and resume information, as well as your desire to work in the specific field. You can mention key personal attributes and accomplishments.

Part 3: List of References

Prepare a list of names and contact information for at least three people who would recommend you for this job. (They will not actually be contacted for this assignment.) Describe your relationship with each person. Make sure the relationship is appropriate for the job you are applying for.

Part 4: Interview

- Prepare a list of 10 interview questions that could be asked by an employer during an interview for this career.
- Prepare a list of five questions that would be appropriate to ask an employer about this career.

Name____ Date____________________

Maria Alvarez

123 Hilldale Ave

Anywhere Town, USA 12345

Phone: 555-242-5752; Cell: 555-545-4813

malvarez@cmail.com

Career Objective: To obtain a position as a veterinary assistant.

Education:

- Springfield High School, Anywhere Town, USA – Diploma, mixed curriculum – June 2023. Courses include introduction to agriculture, animal science, accounting, advanced biology, and chemistry.
- Accepted to Springfield Community College beginning fall 2023.

Skills: Experienced in providing care for animals and children as well as working with the public; familiar with the use of inventory software program in an office setting.

Experience:

Kennel Attendant (2020 – Present)

- Walked dogs, cleaned kennels, and provided customer service.

Child Care (2019 – Present)

- Provide childcare to neighbor children in the summer.

Achievements:

State FFA Degree Recipient

State Finalist in the Small Animal Care Career Development Event

Honor Roll (2021, 2022, 2023)

Volunteer Experience:

Vineyard Food Pantry (2022 – present)

Society for the Prevention of Cruelty to Animals (SPCA) (2022 – present)

Extracurricular Activities:

FFA Leadership Organization (2020 – present)

High School Band (2020–2023)

Youth Group Member (2022 – present)

Varsity Track and Field (2020–2022)

References:

Jim Smith	Agriculture Teacher/FFA Advisor	555-454-4455
Haley Sprout	Employer, Sprout Kennels	555-454-3465
Jamar Kent	Youth Group Leader	555-454-6367

Name______________________________________ Date________________

Exercise 2-4 Fill-in-the-Blank Activity

Directions: Use the word bank below to fill in the blanks in the following statements.

Word Bank:

associate degree
bachelor's degree
doctoral degree
citizenship
master's degree

1. In order to become certified to teach animal science courses at any high school in the country, he went to graduate school and received his ____________________.
2. A local beef farmer showed ____________________ by volunteering a day this summer at an annual local farm tour, giving hayrides to visitors.
3. Before Josh got hired by a dairy equipment company to install equipment into milking parlors, he attended a local college for two years and received his ____________________.
4. She earned her ____________________ in farm business management after attending a state university for four years.
5. A veterinarian has to be successful through the challenges of vet school and residency in order to get their ____________________.

Name_______________________________________ Date____________________

Chapter 2 Lab Questions

1. Describe the occupation trends in agriculture and animal science over the past 25 years.

2. How does the trend of population growth affect animal science jobs?

3. What are the advantages of post-secondary education when seeking a career field?

4. What are the citizenship skills related to a career in agriculture and why are they important? Give an example of each.

Chapter 3

Safety in Livestock Production

Introduction

Agriculture is known to be one of the most dangerous occupations in the United States. Data for agriculture (including farming, forestry, fishing, and hunting) show a death rate of 19.5 deaths per 100,000 workers, while all industries combined have a death rate of about 3.4 deaths per 100,000 workers. Many farm accidents result in nonfatal injuries that can be disabling and life changing. Most nonfatal accidents involve some type of farm machinery, and the second most common agricultural accident involves handling livestock. Each year, approximately 150,000 farm accidents in the United States cause serious injury and disability.

Farm accidents can be costly as well. The annual cost of farm accidents in the United States is about $8 billion in medical costs and lost productivity. To avoid the financial and physical impacts of farm accidents, the law requires that agricultural employers maintain a safe work environment as enforced by the Occupational Safety and Health Act (OSHA). Chemicals are used for many applications in agriculture, and therefore, guidelines are in place to control their use. Policies that govern the use of chemicals in agriculture are created and controlled by the Environmental Protection Agency (EPA). A Safety Data Sheet (SDS) is required to be available for all chemicals in the workplace. The SDS contains the details and characteristics of the chemical and any other hazardous information. Many injuries and even deaths related to improper chemical storage and labeling occur each year.

Given the dangerous nature of the agricultural industry, first-aid preparation is necessary for all farming facilities. First-aid kits should be located in several key areas, such as in the livestock buildings, on the farm equipment, and in all vehicles.

When handing any type of livestock, there is risk of injury to the workers and to the animals. Many accidents occur during loading and unloading of livestock onto trailers. Other hazards include fire, electrical current, manure pits, lagoons, livestock confinement buildings, and grain storage areas. Proper, heavy-duty facilities and equipment help reduce the risk of accidents.

Proper handling techniques and worker awareness will also reduce the number of injuries while handling livestock.

Hazards in Livestock Handling

1. The most common types of animal handling injuries include the following:
   - Animal steps on handler.
   - Animal slips and falls on handler.
   - Animal pins or squeezes handler against a barrier.
   - Animal kicks handler.
2. By employing practical experience and adhering to a few general rules, listed below, handlers can prevent most accidents and injuries.
   - Move calmly, deliberately, and patiently. Avoid quick movements or loud noises that may startle animals.
   - Do not alter the daily routine or the animals' living conditions. Animals often balk at anything out of the ordinary.
   - Always leave an escape route when working in close quarters with animals.
   - Avoid startling an animal. Make sure it is aware of your approach before getting too close to it.
3. Understanding the flight zones of an animal is key to safely moving livestock. Figure 3-9 in the textbook shows the area of sight for cattle. Avoiding the flight zone can help to reduce injuries.

Name________________________________ Date________________

Exercise 3-1 Facilities Designed with Safety in Mind

Using the principles discussed, your task will be to design a safe confinement area to move cows on a local beef farm. The cattle are currently in a large, open pasture, making them hard to corral. The cows need to be corralled and individually restrained to allow the veterinarian to check them for pregnancy. If they are pregnant, they will be released into a brood cow pasture close to the barn; if they are not pregnant, they will be loaded onto a trailer for sale. The farm owner wants a permanent corral system that is ideally designed with cattle behavior in mind as well as handler and animal safety. A map of the current facilities is given here; it includes the current location of the cattle and the desired end location of the brood cow pasture and the trailer.

Your Task: Using the base map of the existing farm (**Figure 3-1**), design a safe corral system, including an effective funnel system to capture cattle from pasture, proper flow for cattle movement, equipment to restrain livestock while veterinarian checks for pregnancy, and proper facilities to separate cattle to pasture or to load onto the trailer. Include worker locations on the map as well.

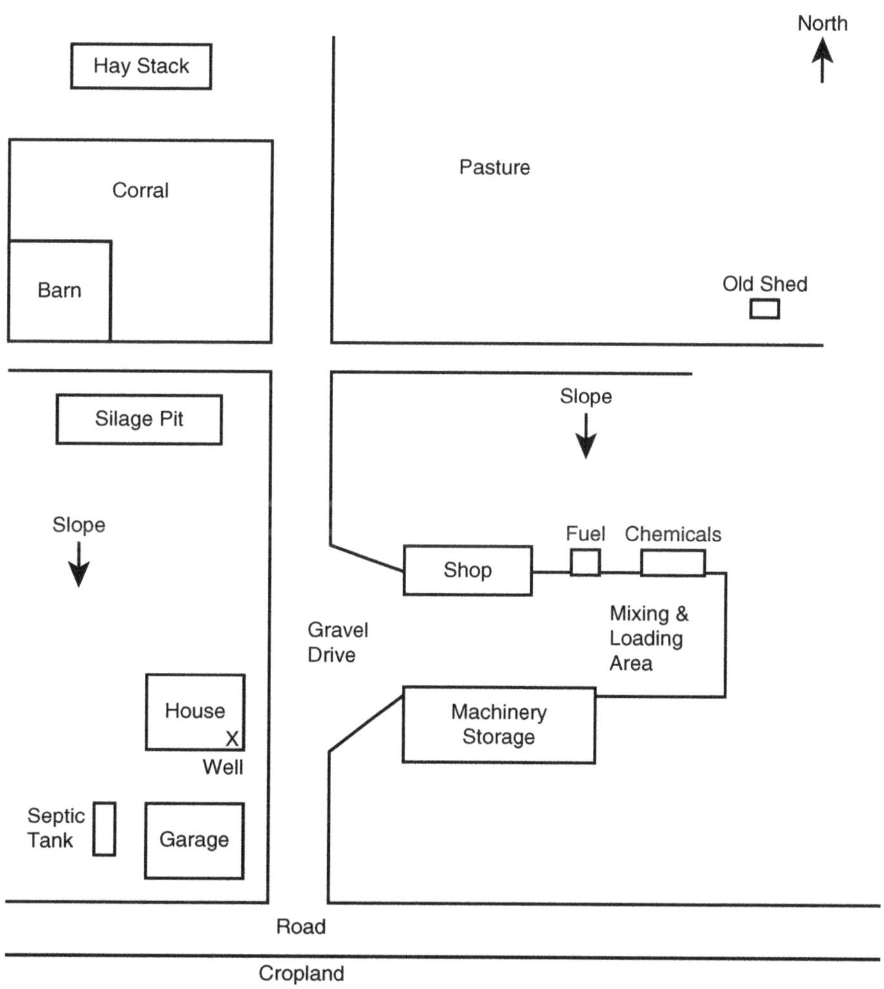

Figure 3-1

Name____ Date____________________

Explain: Describe your facilities and your reasons for designing them this way. What animal behavior and safety principles did you use in your construction?

Name____ Date____________________

Exercise 3-2 Livestock Handling Injury Report

Read the following scenario describing a common accident involving a livestock production facility. After reading the scenario, use the Supervisor's Incident Investigation Report to identify the causes of the accident and recommend changes to prevent similar incidents in the future.

Scenario

On the Chenery farm, a thoroughbred racehorse facility, the local emergency squad received a call for a multiple-victim farm accident. When they arrived at the scene, they found an adult male in his 40s and a 15-year-old teenage female, both of whom had been struck or trampled by two horses that were being loaded onto a trailer. The adult was in the trailer unconscious with lacerations and contusions to the head, as well as several broken bones. The teenager had injuries to the face and legs. The horses were loose and roaming about the farm when medics arrived. Both individuals were airlifted to the closest hospital for critical care.

Once the victims are stabilized and the horses captured and settled down, the investigation must begin. As the supervisor, you must talk to the witnesses and assess the scene to determine what happened. Once the victims are stable, they can be surveyed for information about the accident. In this case, Charlie is a full-time employee who has worked as a groom and hot walker for five years. Charlie is in his 40s and has more than 20 years of experience with horses, but he is sometimes too comfortable with them. Anna Grace is a teenager hired two weeks prior to the incident. Anna Grace has very little experience with horses and does not understand animal behavior very well.

On this day, just around 5:00 a.m., Charlie was loading two thoroughbreds onto a trailer to head to Kentucky. They were running late and needed to get on the road. The loading ramp normally used for getting horses on the trailer was blocked by a tractor with a flat tire, so Charlie decided the quickest alternative was to load the horses onto the trailer from the rear so they would not need the loading ramp.

Anna Grace fetched the horses from their stalls; in a hurry, she slapped on their lead ropes, yelled at the horses to come, and pulled the lead hard as she moved fast toward the trailer. As she got to the trailer holding the leads to both horses, the smooth concrete was slick due to an overnight rain. Both horses lost their footing and stumbled just before stepping on the trailer. The horses reared up and panicked, refusing to get onto the trailer. Anna Grace held onto the leads, but she was losing control, so Charlie took the leads from Anna Grace and tried again to lead the horses onto the trailer. Charlie got both horses right up to the door of the trailer, but they were spooked and refused to enter. Charlie was inside the trailer tugging on the leads. In an effort to move the horses, Anna Grace raised her voice, clapped, and waved her hands from behind to move the horses forward. Both horses lunged forward onto the trailer, pinning Charlie in the trailer; the escape door had never been unlocked. The trailer floor was also slick, so one horse fell to the ground and knocked Charlie down as well; both horses started flailing and trampling Charlie on the trailer floor. Seeing Charlie was in trouble, Anna Grace tried to enter from behind but was kicked in the face by one of the horses

Name______________________________________ Date____________________

and trampled as the horses backed out of the trailer. Both horses ran off to get away from the commotion, and the two workers were left on the ground, injured and helpless. Another worker arrived as the incident happened; she witnessed the injuries, called 911 immediately, and began administering first aid to the victims.

Supervisor's Incident Investigation Report

Directions: Complete this form to describe the incident that resulted in serious injury or illness. Use the scenario to gather information for the report.

This is a report of a: ❑ Death ❑ Lost Time ❑ Dr. Visit Only ❑ First Aid Only ❑ Near Miss	
Date of incident:	This report is made by: ❑ Employee ❑ Supervisor ❑ Team ❑ Final Report

Step 1: Injured employee (complete this part for each injured employee)		
Name:	**Sex:** ❑ **Male** ❑ **Female**	**Age:**
Department:	**Job title at time of incident:**	
Part of body affected: (shade all that apply) 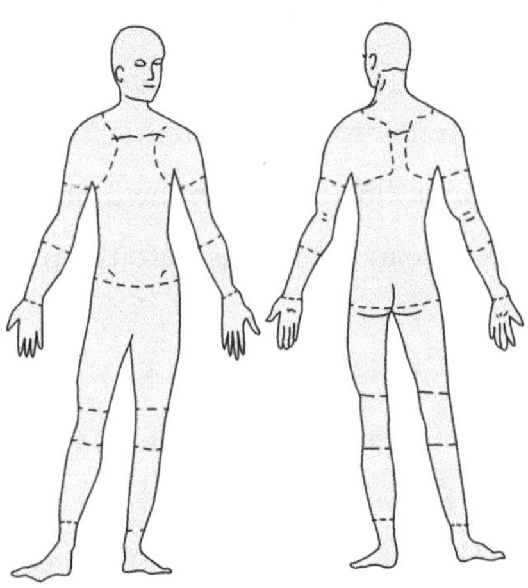	Nature of injury: (most serious one) ❑ Abrasion, scrapes ❑ Amputation ❑ Broken bone ❑ Bruise ❑ Burn (heat) ❑ Burn (chemical) ❑ Concussion (to the head) ❑ Crushing Injury ❑ Cut, laceration, puncture ❑ Hernia ❑ Illness ❑ Sprain, strain ❑ Damage to a body system: ❑ Other ___________	This employee works: ❑ Regular full time ❑ Regular part time ❑ Seasonal ❑ Temporary Months with this employer: Months doing this job:

Name_______________________________________ Date_________________

Step 2: Describe the incident			
Exact location of the incident:			**Exact time:**
What part of employee's workday? ❑ Entering or leaving work ❑ Doing normal work activities ❑ During meal period ❑ During break ❑ Working overtime ❑ Other			
Names of witnesses (if any):			
Number of attachments:	Written witness statements:	Photographs:	Maps/drawings:
What personal protective equipment was being used (if any)?			
Describe, step-by-step, the events that led up to the injury. Include names of any machines, parts, objects, tools, materials, and other important details. Description continued on attached sheets: ❑			

Name____ Date____________________

Step 3: Why did the incident happen?	
Unsafe workplace conditions: (Check all that apply)	Unsafe acts by people: (Check all that apply)
❑ Inadequate guard	❑ Operating without permission
❑ Unguarded hazard	❑ Operating at unsafe speed
❑ Safety device is defective	❑ Servicing equipment that has power to it.
❑ Tool or equipment defective	❑ Making a safety device inoperative
❑ Workstation layout is hazardous	❑ Using defective equipment
❑ Unsafe lighting	❑ Using equipment in an unapproved way
❑ Unsafe ventilation	❑ Unsafe lifting by hand
❑ Lack of needed personal protective equipment	❑ Taking an unsafe position or posture
❑ Lack of appropriate equipment/tools	❑ Distraction, teasing, horseplay
❑ Unsafe clothing	❑ Failure to wear personal protective equipment
❑ No training or insufficient training	❑ Failure to use the available equipment/tools
❑ Other: ____________________	❑ Other: ____________________
Why did the unsafe conditions exist?	
Why did the unsafe acts occur?	
Is there a reward (such as "the job can be done more quickly" or "the product is less likely to be damaged") that may have encouraged the unsafe conditions or acts? ❑ Yes ❑ No If yes, describe:	
Were the unsafe acts or conditions reported prior to the incident? ❑ Yes ❑ No	
Have there been similar incidents or near misses prior to this one? ❑ Yes ❑ No	

Name______________________________ Date______________

Step 4: How can future incidents be prevented?

What changes do you suggest to prevent this injury/near miss from happening again?

❑ Stop this activity ❑ Guard the hazard ❑ Train the employee(s) ❑ Train the supervisor(s)

❑ Redesign task steps ❑ Redesign workstation ❑ Write a new policy/rule ❑ Enforce existing policy

❑ Routinely inspect for the hazard ❑ Provide personal protective equipment

❑ Other: ______________________

What should be (or has been) done to carry out the suggestion(s) checked above?

Description continued on attached sheets: ❑

Step 5: Who completed and reviewed this form? (Please Print)

Written by: Department:	Title: Date:
Names of investigation team members:	
Reviewed by:	Title: Date:

Name______________________________ Date______________

Chapter 3 Lab Questions

1. List five types of agricultural accidents.

2. What are some climatic factors related to safety and how can workers protect themselves?

3. What are five practices that can keep workers safe when handling livestock?

4. Why is it recommended to use curved chutes with solid colored panels for moving livestock?

Feel good studio/Shutterstock.com

Chapter 4

Livestock and the Environment

Introduction

Wastes associated with livestock production have become more of a public concern than in the past. About 2 billion tons of manure are produced each year on livestock farms in the United States. Due to urban sprawl, attention has been drawn to the methods of animal waste removal and the impacts waste disposal has on the environment and the local community. Animal wastes can include manures, soiled bedding products, juices associated with silage feeds, wash-down water from cleaning livestock facilities, manure runoff liquids, disposed animal carcasses, and other types of wastes. To the livestock producer, wastes can be assets if managed properly and can reduce the need for commercial fertilizers that may be harmful to the environment. Wastes also add organic material to the soil improving the fertility of the land.

If not managed properly, animal wastes can have a significant negative effect on the environment. Our natural resources, such as water, soil, and air, may be jeopardized if animal wastes are not managed professionally. Protecting ground and surface water from waste contamination is a priority. Water is usually contaminated by leaching of waste products through the ground or by wastes running along the surface of the ground into a water source. The worst offence of water contamination is direct discharge into lakes and streams.

In livestock husbandry, waste accumulates every day. To manage these wastes properly to protect the environment and surrounding community, a livestock producer must develop a waste handling plan. A waste handling plan should include:

1. Information from area soil surveys.
2. A current nutrient analysis for all areas on which waste will be used, including anticipated crops to be grown.
3. A map of the farm showing roads, wells, waterways and basins, dwellings, and waste treatment and holding facilities.
4. Types and amounts of wastes and their nutrient values.
5. Rates and times of application.

In this chapter, you will learn to understand the environmental laws affecting the management of animal wastes. You will determine what the specific laws are in your state and how these environmental laws affect an animal producer. Next, you will calculate the amounts of wastes that are produced by different types of livestock and calculate the amount of manure needed to fertilize a given amount of field area. You will also create a map of an example farm, keeping waste handling considerations in mind.

istock.com/Hans Verburg

Figure 4-1 Manure being spread onto crop fields as fertilizer.

Name____ Date__________________

Exercise 4-1 Environmental Impact of Wastes (Federal and State Laws)

Directions: For each of the following environmental laws, answer these questions: (1) How does the law protect the environment? (2) How does this limit or affect farming practices? (3) Which government agency enforces this law?

The Federal Water Quality Act of 1965

1. ____________________
2. ____________________
3. ____________________

The Refuse Act of 1899

1. ____________________
2. ____________________
3. ____________________

The Solid Waste Disposal Act of 1965

1. ____________________
2. ____________________
3. ____________________

The Federal Clean Air Act

1. ____________________
2. ____________________
3. ____________________

Your State Laws

1. ____________________
2. ____________________
3. ____________________

Name____ Date____________________

Exercise 4-2 Manure Application Calculations

When applying wastes to the land, the farmer must know how much waste is produced each day, as well how much to apply to various areas of crop land.

Directions: Use the following scenario and values to answer the questions about waste application rates. Use Tables 4-1 through 4-3 in the textbook to find the values needed in the calculations.

Scenario

Joe is the crop specialist for a large dairy operation. One of his responsibilities is to manage the waste handling operation at the farm. As the manure is gathered at the facilities, Joe has a team of operators spreading the wastes onto their crop fields to fertilize the land. In this dairy operation, they milk 1,200 dairy cows every day and must remove all the waste from the barns. Joe has an expected corn yield of 170 bushels per acre. The livestock waste to be applied comes from pit storage that Joe uses to contain manure when conditions are not favorable for spreading it on fields. Joe has tested the nitrogen content of the manure and found it to be 40 pounds of nitrogen per 1,000 gallons of manure. When the time is right, Joe will apply the manure on the fields. He has 700 acres of land to fertilize with this manure.

Calculation Questions

1. How many pounds of wastes are produced at Joe's farm each day?

____

____

2. How many pounds of wastes are produced over the entire year (365 days)?

____

____

3. How many pounds of nitrogen are needed per acre?

____

____

4. How many gallons of waste will need to be applied per acre to grow the corn?

____

____

Name____ Date__________________

5. How many total gallons of waste will Joe need to apply to cover all the farm property?

6. Will Joe's farm produce enough waste to properly fertilize the total farm property?

7. If Joe doesn't produce enough waste to fertilize his own farm, what are some alternatives?

Name_______________________________________ Date__________________

Chapter 4 Matching Activity

Directions: Match the term to the corresponding definition or description.

Term

____ 1. point source
____ 2. agronomic nitrogen rate
____ 3. diversion
____ 4. drainage channel
____ 5. debris basins
____ 6. holding pond
____ 7. disposal
____ 8. masking agents
____ 9. digestive deodorants
____ 10. effective ambient temperature (EAT)
____ 11. diverse terrace
____ 12. counteractants
____ 13. deodorants
____ 14. comfort zone
____ 15. lower critical temperature (LCT)
____ 16. upper critical temperature (UCT)
____ 17. wind chill index
____ 18. estray

Definition

a. collects runoff water from a feedlot
b. final step in controlling runoff from feedlots; the collected water can be used for irrigation of the land
c. bacteria that create a digestive process that eliminates odor
d. temporary storage area for runoff; it is not designed for waste treatment
e. cover up the odor of wastes with the introduction of another odor
f. a large source of pollution; the law prohibits discharge of pollutants from a point source into a river or stream without a permit
g. combined effect of factors such as humidity, precipitation, wind, and heat radiation on the efficiency of energy use by farm animals
h. preventing surface water from outside the feedlot from getting onto the feedlot
i. keeps about 50 to 85 percent of the solids from reaching the holding ponds
j. amount of available nitrogen per unit of yield necessary to produce a given crop
k. domestic animal of unknown ownership that is running at large
l. temperature at which animals will show symptoms of heat stress; feed intake is generally lower as animals attempt to reduce the rate of metabolic heat production when the upper critical temperature is reached
m. forces the water to go around the feedlot
n. chemicals that kill bacteria that cause odor
o. measure of the combined effect of air temperature and speed of air movement
p. attempt to neutralize odor
q. also called the thermoneutral zone; the range of effective ambient temperatures within which an animal does not have to increase normal metabolic heat production to offset heat loss to the environment
r. temperature at which animals will show symptoms of cold stress; feed intake increases, as does metabolic heat production

Name______________________________ Date______________

Chapter 4 Lab Questions

1. How much manure is produced on livestock operations in the United States every year? What is causing the increase in the concentration of animal wastes on individual farms?

2. What are the factors to consider when choosing a manure handling system?

3. Why is manure an environmental concern? What are the possible environmental impacts?

4. On a separate page, develop a waste handling plan for a local livestock producer.

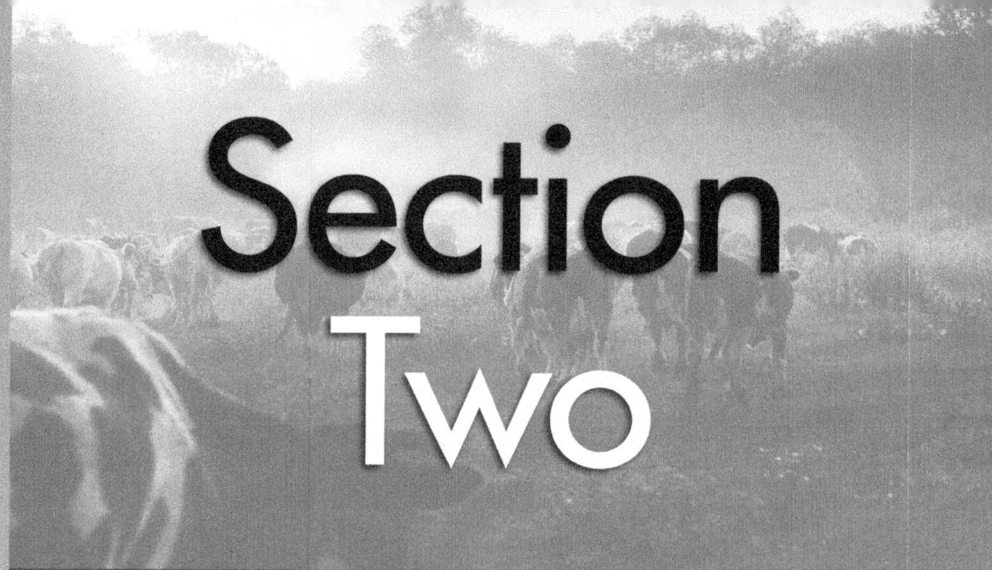

Section Two

Anatomy, Physiology, Feeding, and Nutrition

Feel good studio/Shutterstock.com

Chapter 5

Anatomy, Physiology, and Absorption of Nutrients

Introduction

It is important that professionals in the animal industry have a working knowledge of animal anatomy and physiology. Such knowledge is essential when making management decisions and implementing health care treatment. Decisions affecting feeding strategies, shelter, reproductive programs, and much more depend on the knowledge of those managing the animals. For example, a farmhand at a livestock operation must be able to identify abnormalities, injuries, or even a change in behavior and/or appearance. If the farm employee can identify the problem, they can notify the farm owner and treatment can begin. The owner and veterinarian can be called in to give a diagnosis and prepare a treatment strategy. Even owners of companion animals should have a knowledge of basic animal anatomy to recognize health issues and get treatment.

Many kinds of tissues make up the bodies of animals. Tissues are groups of specialized cells that form individual organs, as well as complete systems that carry out the functions of the body. The animal body can be classified into 10 major systems: skeletal, muscular, respiratory, circulatory, integumentary, immune, nervous, renal, reproductive, and digestive. Refer to the textbook chapter for detailed information on each system.

This chapter will explore each of the body systems through an anatomy model activity as well as a research activity involving animal hormones. Chapter matching and vocabulary will help develop an understanding of anatomy terms.

Name____ Date_____________________

Exercise 5-1 Anatomy Model Lab

The purpose of this exercise is to identify the internal parts of a horse. The instructor will provide you with materials to create a model.

1. *Directions:* Using the following diagram (**Figure 5-1**) and other sources for reference, construct a horse model. All internal parts of the horse must be labeled. When the model is complete and labeled, the following three criteria will be used to grade the model:
   - Accuracy and location of all internal parts of the anatomy
   - Construction quality (clean, neat, etc.)
   - Labeling (accurate and legible)

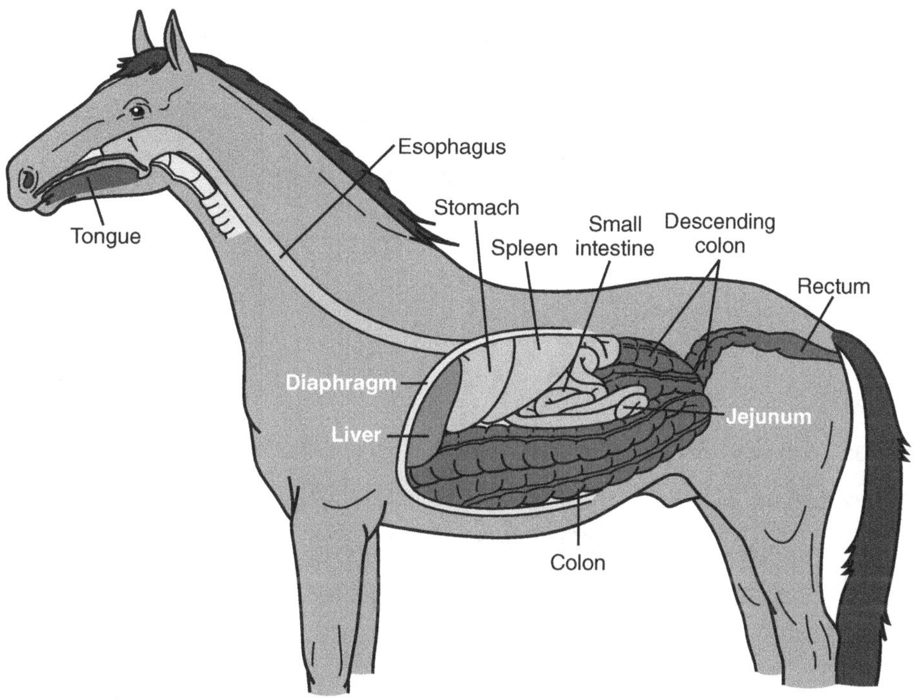

Figure 5-1 Internal anatomy of a horse.

Name____ Date____________________

Extension Activity

Paint a skeleton on the coat of a well-behaved horse. Use paint that is suitable for application on animals and easy to wash out. Perform a demonstration to a young group of students that teaches the shapes and names of the parts of the skeletal system. You may use **Figure 5-2** or other sources for reference. When finished, wash the paint off the horse.

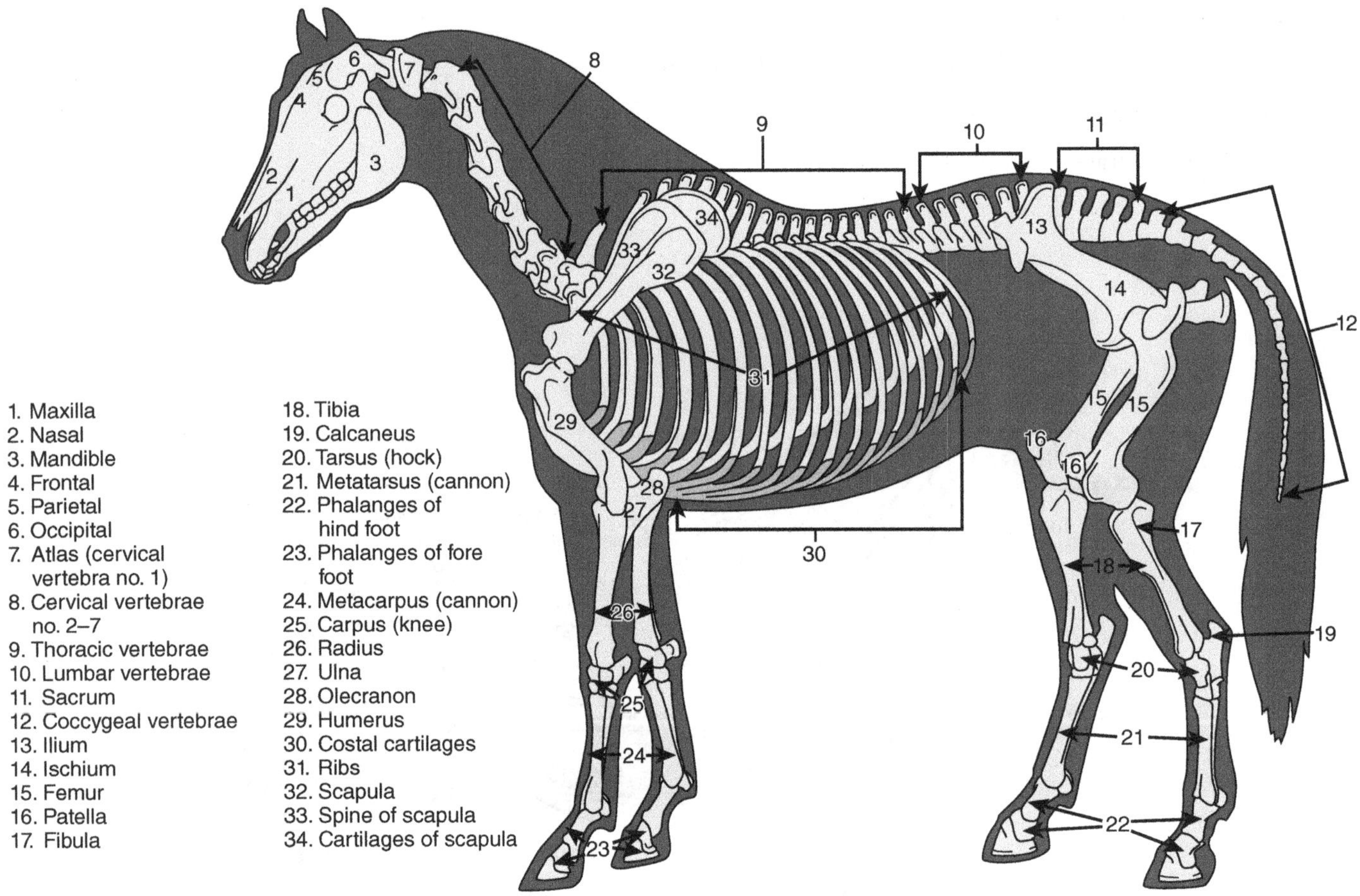

Figure 5-2 Labeled horse skeleton.

Name________________________________ Date______________

Exercise 5-2 Functions of Hormones

The endocrine system is reliant on hormones to act as chemical messengers to maintain the growth and development of the body. The glands of the endocrine system produce specific hormones, and as blood passes through, the gland absorbs the hormone and transports it to its target organ or tissue.

Directions: For each of the hormones listed, (a) describe the function of each hormone, and (b) identify which gland produces the hormone.

1. Aldosterone

   a. ______________________________

   ____

   b. ______________________________

   ____

2. Epinephrine, also called adrenaline

   a. ______________________________

   ____

   b. ______________________________

   ____

3. Luteinizing hormone-releasing hormone (LHRH)

   a. ______________________________

   ____

   b. ______________________________

   ____

4. Estrogen

   a. ______________________________

   ____

   b. ______________________________

   ____

Name______________________________ Date________________

5. Progesterone
   a.
   b.
6. Insulin
   a.
   b.
7. Follicle-stimulating hormone (FSH)
   a.
   b.
8. Gonadotropin
   a.
   b.
9. Luteinizing hormone (LH)
   a.
   b.

Name______________________________ Date____________

10. Somatotropin

    a.

    ____

    ____

    b.

    ____

    ____

11. Oxytocin

    a.

    ____

    ____

    b.

    ____

    ____

12. Testosterone

    a.

    ____

    ____

    b.

    ____

    ____

Name____ Date____________________

Chapter 5 Matching Activity

Term	Definition
____ 1. exoskeleton	**a.** how cells divide and grow and differentiate into tissues with a variety of functions
____ 2. fibrous joints	**b.** a group of organs that carries out a major function
____ 3. amphiarthroses joints	**c.** a gelatin-like protein found in bones
____ 4. cells	**d.** specialized cells that break down the cartilage and replace it with bone
____ 5. diarthroses	**e.** typically, a hard shell on the outside of the body
____ 6. synovial membrane	**f.** found only in the muscular wall (myocardium) of the heart; it is striated in the same manner as skeletal muscle
____ 7. striated voluntary muscle	**g.** flesh that has dark bands that cross each muscle fiber
____ 8. unstriated muscle	**h.** where the actual exchange of gases occurs; the walls are very thin, about one cell thick, and are covered by a film of fluid that acts as a surfactant
____ 9. system	**i.** carries blood through the lungs of the animal
____ 10. cardiac muscle	**j.** connects to the capillary bed surrounding the alveoli
____ 11. larynx	**k.** joined by fibrous tissue or, in some cases, cartilage tissue; these joints do not permit any type of movement
____ 12. trachea	**l.** a tube that leads to the bronchi; the wall is lined with a series of C-shaped rings of cartilage
____ 13. alveoli	**m.** consist of discs of a fibrous cartilage that separate and cushion the vertebrae, allowing very limited movement
____ 14. arterioles	**n.** an area composed of cartilage structures; it contains vocal cords that vibrate when air passes across them, producing sound
____ 15. pulmonary circulation system	**o.** refers to taking the digested parts of the feed into the bloodstream
____ 16. ossein	**p.** joints that allow free movement and have a fluid-filled cavity
____ 17. plasma	**q.** when ball-like masses are formed in the stomach and the material is then forced back up the esophagus to be chewed again
____ 18. rumination	**r.** muscles that surround the hollow internal organs of the body, such as the blood vessels, stomach, intestines, and bladder
____ 19. absorption	**s.** the fluid portion of whole blood
____ 20. osteocytes	**t.** usually found inside the ligaments that help hold the joint together

Name______________________________ Date______________

Chapter 5 Lab Questions

1. What are the main functions of the skeletal system?

2. Name and describe the three types of muscle normally found in animal bodies.

3. How does the avian respiratory system differ from the mammalian system?

4. What is the importance of the endocrine system?

5. Why can ruminants digest large amounts of roughage?

Feel good studio/Shutterstock.com

Chapter 6

Feed Nutrients

Introduction

Domesticated animals rely on humans to provide feed that contains adequate nutrients to sustain various stages of growth and development. A nutrient is a chemical that aids in the support of life and provides energy by being absorbed into the body cells. Nutrients are divided into five groups: (1) energy nutrients (carbohydrates, fats, and oils), (2) proteins, (3) vitamins, (4) minerals, and (5) water. Animals require many different nutrients to maintain the daily functions of the body. Each nutrient needs to be provided to the animal in balance with the other nutrients. To provide proper nutrition, an animal nutritionist must create a feed that has balanced rations of each nutrient. These rations are calculated using percentages, and they can be adjusted to provide different levels of nutrients, depending on the animals' stages of growth. Too much of one nutrient and not enough of another may result in unhealthy animals, low production, and high feed costs.

Animals differ in the kinds and amounts of nutrients they require. A lactating dairy cow has a higher nutrient requirement than a retired racehorse. It is the job of animal nutritionists to determine which nutrients animals need and then make recommendations for feeding each kind of animal. The recommendations are based on the results of feeding experiments and nutrient sampling and testing, as well as nutrient content calculations. A good nutritionist will recommend a feeding strategy that provides proper nutrition for growth and production, while also being economical.

Small-scale animal producers or hobby farmers may purchase bagged feed from a grain elevator or a local feed store. When purchasing bagged feed, a producer should understand the information on a feed tag in order to purchase the correct feed for their animals. On the tag, the maximum amount of available fiber in that feed is shown as a percentage of the total weight of the bag of feed. A list of ingredients will also be printed on the tag to show the sources of carbohydrates contained. Some common examples of carbohydrates in bagged feed are cane molasses, ground corn, wheat middlings, and oats.

In this chapter, you will conduct a feed sampling activity as well as map the path that grain travels from the fields where the grain is produced to the grain elevator and eventually to the farmer.

Name____ Date____________________

Exercise 6-1 Feed Sampling Activity

Accurate feed analyses are necessary for animal producers to balance feeding rations, accurately price hay, formulate needed supplements, and determine the nutrient value of feeds made on the farm. Adequate forage testing is necessary to develop an accurate feed inventory. An inventory is necessary to allocate the high-nutrient feeds to high-producing livestock and the lower-nutrient feed to animals at a lower level of nutrient requirements.

Directions: Using the sampling guidelines page, sample various "lots" of feed. Access to feed inventories will be made available by the instructor. Accurately obtain the samples and prepare them for testing. For each type of feed, record the date and time, identify the "lot" of feed, and describe how the sample was prepared and handled in detail.

Feed Sampling Report

Test 1: Round bales

Date and time: ____________________

Lot of feed: ____________________

Sample preparation/handling details:

____

____

Test 2: Compressed loaf haystack

Date and time: ____________________

Lot of feed: ____________________

Sample preparation/handling details:

____

____

Test 3: Chopped or ground hay

Date and time: ____________________

Lot of feed: ____________________

Sample preparation/handling details:

____

____

Test 4: Harvest silage (fresh as it comes out of the field)

Date and time: ____________________

Lot of feed: ____________________

Sample preparation/handling details:

____

____

Name____ Date____________________

Test 5: Silage at feeding from an upright silo

Date and time: ____

Lot of feed: ____

Sample preparation/handling details:

____

____

Test 6: Silage from horizontal silo

Date and time: ____

Lot of feed: ____

Sample preparation/handling details:

____

____

Sampling Guidelines

The feed values of most forages vary due to factors such as time of harvest, soil quality, moisture content, and many other factors. To accurately manage a feed inventory, test forages routinely to determine when and how the feed should be used. An accurate forage inventory is necessary so the producer can allocate high-protein feeds to animals with a higher nutrient demand and allocate slightly poorer quality feeds to livestock that have a low nutrient requirement. This guideline focuses on the sampling strategies when testing hay and silage.

The Sample "Lot"

Samples must closely resemble the entire lot of forage. Each sample should only represent one lot. A *lot* is a term used to describe forage that was all harvested from the same field within a 48-hour period and usually contains less than 100 tons of forage. The most important characteristic of a lot is that it is uniform. All forage from the same lot should have following similarities: type of plant(s), field soil type, cutting date, maturity, variety, weed contamination, type of harvest equipment, weather during growth and harvest, preservatives, drying agents, additives, storage conditions, and pest or disease damage. The consistency of these factors is essential to obtaining an accurate sample. If any of these factors differ, designate a new lot of forage.

Hay Sampling

Sample baled long hay after curing. A core sampler or a probe is an essential tool for sampling hay. The core sampler should be inserted into the bale at least 12 to 18 inches and have and internal diameter of at least ⅜ inch. Core samplers usually have sharpened or serrated tips to cut through hay. Several core samplers are designed to be operated with an electric hand drill. The sampler should be inserted at a slow speed to avoid the sample heating up and then being unfit for testing.

Name________________________________ Date________________

Square or Round Bales

Select 20 or more average bales from each lot. Collect one sample from each bale by coring the square bale straight in from the center of the end of the bale. Take samples from any broken hay bale piles or haystacks that will be used to feed livestock in some way.

Loosed or Compressed Haystacks

When collecting samples, use a hay probe at least 24 inches long to collect 15 or more samples from each lot. When sampling loose haystacks, make sure to take core samples from six general locations: (1) top front, (2) top middle, (3) top rear, (4) lower front side, (5) lower middle side, and (6) lower rear side. When top locations are sampled, stand on the top of the stack and insert the probe vertically downward. For side sampling, use a light downward angle with the probe to avoid sampling parallel stems in the stack.

Chopped or Ground Hay

On a regular basis, collect 10 samples from each lot of hay during grinding, and place the samples in a sealed plastic bag. When sampling previously ground hay in a pile, collect about one-quarter of the samples from the top half of the pile and the rest from the lower half.

Silage Sampling

Silage can be sampled when entering the silo, when feeding after storage, or a combination of both. When forage is properly ensiled, results from fresh samples will agree closely with fermented forage. Avoid collecting samples from rotted or poorly preserved materials in the silage; this kind of spoilage is usually found at the top of upright silos or the shallow end and slopes of a bunk silo.

Fresh Samples

Collect 20 or more samples from each lot of fresh silage as it comes out of the field. Take periodic samples as loads are brought to silo while making sure the container remains sealed to reduce loss of moisture content. For example, when sampling a large silo, collecting one quart from each load in the morning, at noon, and in the afternoon will give an accurate sample of the hay inventory. The following procedure will help ensure an accurate sampling method:

1. Collect at least two gallons of samples and mix well.
2. Make a cone-shaped pile of the forage sample.

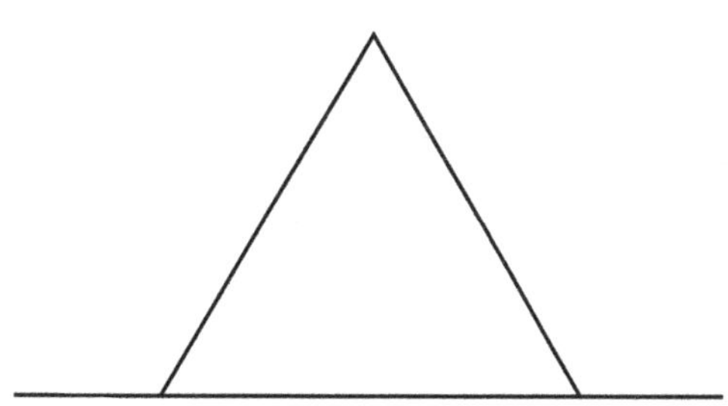

Name____ Date_____________________

3. Divide the pile as if dividing a pie into four pieces.

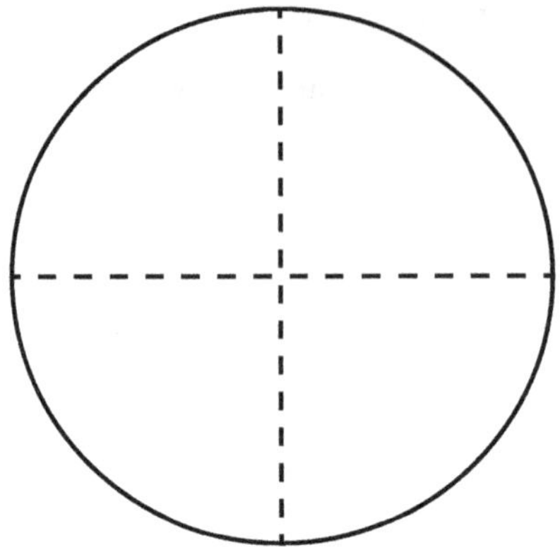

4. Randomly choose one of the quarters to keep and then choose the diagonally positioned quarter; discard the other two quarters.

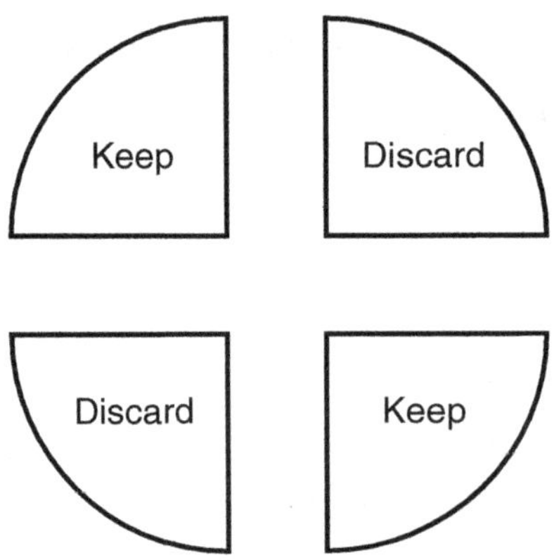

5. Repeat this procedure until 1 to 1½ quarts remain. Then transfer all materials into a two-quart plastic freezer bag. Remove the air and seal tight.

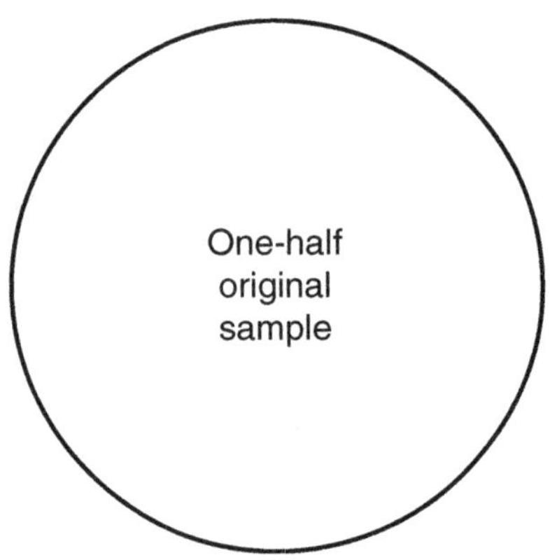

Name____ Date____________________

Upright Silos at Feeding

Collect two to three gallons of feed samples in two-quart increments by passing a clean container under the chute while unloading ensilaged materials. Another method is to collect 20 handfuls of silage as it is fed out to livestock in different locations of the feed bunk. Mix and subsample, as described in the preceding five sampling steps.

Horizontal Silos

Collect 20 or more hand samples from various locations of the exposed face of the silo to represent the entire exposed surface. Mix and subsample as described earlier.

Name____ Date____________________

Exercise 6-2 Grain Elevator Map

Grain elevators are a key component of the current livestock industry in the United States, and the elevator manager must ensure that all the feed company's clients get a high-quality feed with no mold or other contaminants. In this activity, assume the role of the grain elevator manager. The feed company has decided to expand the elevator to accommodate all its clients' needs. The elevator will need to be equipped with enough silos and augers to provide all the needed feed. The other responsibility will be to design the layout of the elevator to allow each silo to unload onto railroad cars.

Directions: Using the sample grain elevator maps for reference (**Figures 6-1** and **6-2**), design a grain elevator operation that can accommodate the needs of the feed company's clients. All the grain silos must be able to unload onto a railroad car. Use technical drawing paper, a straight edge, and a pencil to create the elevator map.

Client Needs

Feed	Processing Procedure	Purpose of Processing
Barley (whole)		
Cracked shelled corn	Shelled, cracked (dry rolled)	Increase digestibility
Crimped oats	Crimped (steam rolled)	Increase digestibility
Dried beet pulp	Dried by-product of sugar production	Ease of handling
Ground shelled corn	Shelled, ground (dry rolled)	Increase digestibility
Liquid molasses	By-product of sugar production	
Shelled (whole kernel) corn	Shelled (removed from cob)	Increase nutrient concentration
Soybean meal	By-product of oil extraction	
Steam rolled oats	Steam rolled	Increase digestibility
Wheat middlings	By-product of flour milling	

© B Brown/Shutterstock.com

Figure 6-1 Grain elevators.

Name_______________________________ Date_______________

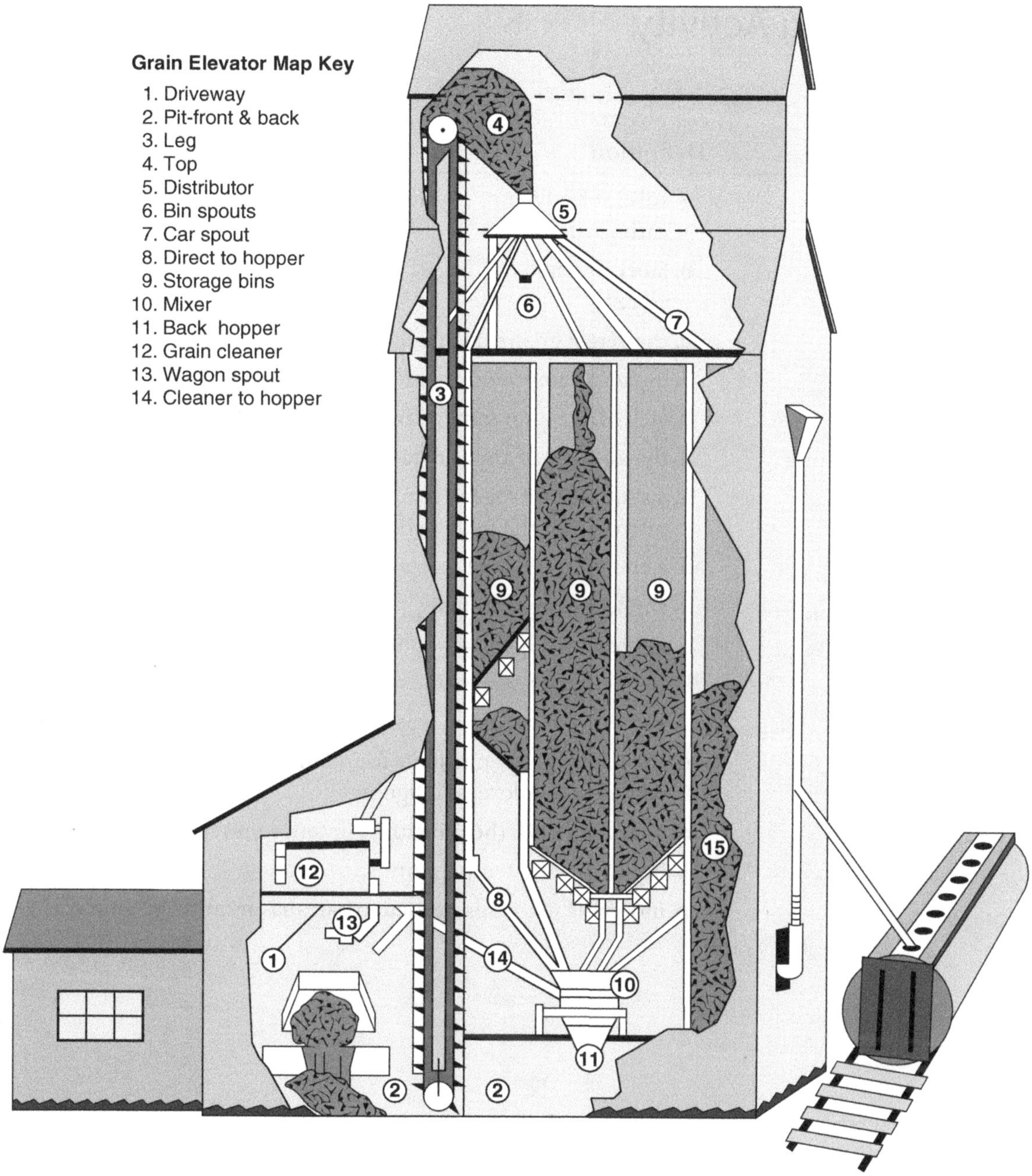

Figure 6-2 Map of parts of a grain elevator.

Name______________________________ Date______________

Chapter 6 Matching Activity

Directions: Match the term to the corresponding definition or description.

Term

____ 1. nutrient
____ 2. carbohydrate
____ 3. fiber
____ 4. commercial feed tag
____ 5. fat
____ 6. oil
____ 7. protein
____ 8. crude protein
____ 9. digestible protein
____ 10. deficiency
____ 11. nitrogen-free extract
____ 12. urea
____ 13. monogastric
____ 14. vitamin
____ 15. mineral
____ 16. dry weight

Definition

a. solid at body temperature; may contain a higher energy value than carbohydrates
b. label attached to a bag of feed purchased at a grain elevator or feed store
c. the cereal grains, such as corn, wheat, oats, barley, rye, and sorghum
d. organic compounds made up of amino acids
e. lack of a certain mineral in an animal's ration
f. the amount of ammoniacal nitrogen in the feed multiplied by 6.25
g. a synthetic nitrogen source that is mixed in the ration to provide nitrogen for the making of amino acids in the rumen
h. a chemical element or compound that aids in the support of life
i. single-stomach animal
j. approximately the amount of true protein in a feed
k. trace organic compounds needed only in small amounts by animals
l. substances that are more difficult to digest than simple carbohydrates
m. the main energy nutrients found in animal rations; made up of sugars, starches, cellulose, and lignin
n. its weight with the moisture content removed
o. liquid at body temperature
p. inorganic materials needed in various amounts by animals

Name____________________ Date__________

Chapter 6 Lab Questions

1. What is a nutrient?

2. List the six groups of nutrients.

3. What is a balanced ration?

4. What factors will make feed rations differ from each other?

Feel good studio/Shutterstock.com

Chapter 7

Feed Additives and Growth Promotants

Introduction

Feed additives are materials used in an animal feed ration that are not classified as nutrients. These materials are used in small amounts to improve feed efficiency, promote faster growth gains, improve animal health, or increase production of animal products. Adding antibiotics at low levels in the ration over a period of time is a common practice in livestock feeding. Many different additives are available, and each has a different desired effect on livestock performance. Too much of one type of additive can cause negative health effects.

Antibiotics are compounds used to kill or slow the growth of some kinds of microorganisms. Other additives such as dewormers (anthelmintics) can be used to control various species of worms that can infest animals. Each species of animal has different needs and not every additive is approved for all livestock. Beef producers commonly use hormone implants; a high percentage of beef cattle rations use either monensin sodium (Rumensin) or lasalocid sodium (Bovatec) to increase the daily rate of weight gain. Feed additives are not as common in dairy cattle because their milk cannot be contaminated by hormones or antibiotics. Young dairy cattle that are not producing milk commonly receive a medicated feed until they near the age of lactation. There are very few additives that have been approved for mixture in sheep and goat feed rations. In swine feed, additives are a common practice to reduce mortality, especially in young stock. Some poultry producers also use medicated feeds to improve growth rates in broilers and egg production in layers. In horses, the use of feed additives is not common, except for occasional use in young colts and fillies.

The Food and Drug Administration (FDA) regulates the use of feed additives, including hormones and antibiotics. There has been heightened concern about the long-term health hazards associated with the use of medicated feeds. The microorganisms' resistance to antibiotics causes the antibiotics to lose their effectiveness against common livestock sicknesses. This chapter explores additives through a feed additives analysis and an antibiotic residue exercise.

Name____ Date_____________________

Exercise 7-1 Feed Additive Analysis

Directions: For the following descriptions of feed additives and their functions, (1) identify each additive, (2) identify which classification of additive it belongs in, and (3) describe how and why the additive has such an effect on the livestock.

Description 1: A beef farmer uses Bovatec trying to increase his cattle's rate of gain.

1. ____________________
2. ____________________
3. ____________________

Description 2: MGA is added to a heifer feed for a growing group of dairy heifers.

1. ____________________
2. ____________________
3. ____________________

Description 3: Implants are used on feeder cattle to increase the proportion of lean meat deposition on the carcass.

1. ____________________
2. ____________________
3. ____________________

Description 4: Monensin is used to improve feed efficiency and rate of gain in replacement dairy heifers.

1. ____________________
2. ____________________
3. ____________________

Description 5: Anhydrous ammonia (NH_3) is being used as a forage preservative for poor-quality forage.

1. ____________________
2. ____________________
3. ____________________

Name____ Date____________________

Description 6: Sodium bicarbonate (backing soda) is used on a group of dairy cattle for four months after calving.

1. ____
2. ____
3. ____

Description 7: Lactating dairy cows receive isoacids for 12 weeks while in high milk production.

1. ____
2. ____
3. ____

Name______________________________ Date______________

Exercise 7-2 Antibiotic Residues

Feed additives containing medications can be an economical method for keeping animals healthy during growth. Proper management of medicated feeds is important to avoid sending animals to the butcher that may still have antibiotic residues in their bloodstream. When caught by U.S. Department of Agriculture (USDA) testing, antibiotic residues will cause livestock to be rejected from slaughtering facilities. In this activity, answer questions based on a pig operation raising suckling pigs. The pigs are farrowed on April 2 and will be processed at a USDA-inspected facility on May 21.

Included below is a sample Medicated Feed Label for swine.

Sample Medicated Feed Label

SuperStart

Medicated

For control of porcine colibacillosis (weaning pig scours) caused by susceptible strains of *ESCHERICHIA COLI.* Carefully follow the feeding directions and WARNING statement printed on the back of this label.

Active Drug Ingredient: Apramycin (as apramycin sulfate)..................150 grams per ton

Guaranteed Analysis

Crude Protein, not less than 21%
Crude Fat, not less than 10%
Crude Fiber, not more than 2.50%
Vitamin A, I. Units per lb. (min) 6,000
Vitamin D3, I. Units per lb. (min) 750.0
Vitamin E, I. Units per lb. (min) 55.0
Riboflavin, mg per lb. (min) 4.5
Niacin, mg per lb. (min) 30.0
d-Pantothenic Acid, mg per lb. (min) 15.0
Choline, mg per lb. (min) 550.0
Vitamin B12, mg per lb. (min) 0.022
Menadione (Vitamin K), mg per lb. (min) 4.5
Biotin, mg per lb. (min) 0.09
Folic Acid, mg per lb. (min) 0.09
Pyridoxine, mg per lb. (min) 0.018
Thiamine, mg per lb. (min) 0.009
Lysine, not less than 1.60%

INGREDIENTS: Dried skim milk, dried whey, animal plasma, heat processed soybeans, fish meal, feeding oat meal, ground corn, meat and bone meal, corn distillers dried solubles, natural and artificial flavors added, sugar, yucca schidigera extract, dehydrated yeast culture, animal fat, cane molasses, monosodium glutamate, methionine, lysine, vitamin A acetate, D-activated animal sterol (source of vitamin D3), riboflavin supplement, niacin supplement, calcium panthothenate, choline chloride, vitamin B12 supplement, menadione dimethylpyrimidionol bisulfite (source of vitamin K), dl alpha tocopheryl acetate (source of vitamin E), biotin, folic acid, pyridoxine hydrochloride, thiamine mononitrate, calcium carbonate, salt, dicalcium phosphate, magnesium oxide, manganous oxide, ferrous sulphate, copper sulfate, cobalt carbonate, ethylenediamine dihydriodide, zinc sulphate, and sodium selenite. **SuperGrow Feed Co. • Toledo, Iowa 52342**

Name____ Date____________________

Feeding Directions: SuperStart AP-150, medicated is a highly palatable product formulated especially for baby pigs being weaned at three weeks of age or earlier and a special formulation for "tail enders" needing a nutritional boost.

Begin feeding SuperStart AP-150, medicated, when pigs are approximately 5 lbs. body weight (1 week of age) and feed continuously and as the sole ration until the pigs have consumed at least 5 lbs. per pig or at least 7 to 10 days after weaning. NEVER WEAN AND CHANGE FEED SOURCE AT THE SAME TIME.

For "tail enders," separate pigs according to size. It is recommended to group pigs by size and place them in groups of 20 or less with a weight difference of no more than 10%.

Feed SuperStart AP-150, medicated, continuously and as the sole ration for at least 3 to 4 weeks or until pigs regain their healthy bloom. SuperStart can also be used as a high nutrient dense product in any starting program to encourage early dry diet consumption.

NOTE: Strains of organisms vary in their degree of susceptibility to antibiotics. If no improvement is observed after the recommended treatment, diagnosis and susceptibility should be reconfirmed.

IMPORTANT: Store in a clean, dry area, free of all offensive odors.

Warning:

DISCONTINUE USE OF THIS MEDICATED FEED 28 DAYS BEFORE SWINE ARE MARKETED FOR HUMAN CONSUMPTION

Directions: Study the feed label above and answer the following questions based on the information in the label:

1. What is the name of the feed?

____

____

____

2. Who manufactured it?

____

____

____

3. What size of animals should it be fed to?

____

____

____

4. What form is the feed?

____

____

____

Name______________________________ Date_______________

5. What ingredients are in this feed?

6. What ingredient is in the largest quantity?

7. Is this a medicated feed?

    - What medication(s) is(are) in the feed?
    - What is the withdrawal time for animals on this medicated feed?

8. How much protein is in this feed?

9. How many pounds are in the final feed mix?

10. What can you tell by the order in which the ingredients are listed?

11. What are the feeding directions?

Name____ Date____________________

12. What cautions are listed?

____

____

____

13. For the suckling pigs in this example, when should the medicated feed be started? When should it be stopped?

____

____

____

Name______________________________ Date______________

Chapter 7 Matching Activity

Directions: Match the term to the corresponding definition or description.

Term	Definition
____ 1. feed additive	**a.** the use of medicated feeds at a lower level to treat sick animals
____ 2. growth promotant	**b.** a compound used to combat a problem that is not susceptible to an antibiotic or antibacterial individually
____ 3. antibiotic	**c.** the period of time necessary for an antibiotic's residues to be eliminated from the bloodstream
____ 4. antimicrobial	**d.** the substance allows animals to use available nutrients more effectively
____ 5. subtherapeutic	**e.** are present in the animal's body at levels too low to cause visible effects
____ 6. subclinical disease	**f.** are used to treat for worms in swine
____ 7. hormone	**g.** materials used in animal rations to improve feed efficiency, promote faster gains, improve animal health, or increase production of animal products
____ 8. anthelmintic	**h.** suppresses estrus (prevents the heifer from coming into heat), which reduces the continual mounting seen when heifers are coming into heat in the feedlot
____ 9. melengestrol acetate (MGA)	**i.** added to cattle, sheep, swine, and poultry rations to prevent coccidiosis
____ 10. withdrawal period	**j.** hormone implants
____ 11. chemobiotic	**k.** compounds such as yeast and lactobacilli that change the bacterial population in the digestive tract to a more desirable type
____ 12. nutrient sparing	**l.** kill or slow down the growth of some kinds of microorganisms
____ 13. coccidiostat	**m.** refers to both natural hormones and synthetic hormone-like compounds
____ 14. probiotic	**n.** antimicrobial compound

Name________________________________ Date________________

Chapter 7 Lab Questions

1. What are the benefits of feed additives?

2. List examples of various additives used in livestock feeds.

3. In which livestock industry is the use of feed additives most common? Why?

4. Why is there concern over long-term use of antibiotic additives in feeds?

Feel good studio/Shutterstock.com

Chapter 8

Balancing Rations

Introduction

Providing an appropriate, balanced ration to livestock is the key to efficient production. The combination of concentrates, feed additives, and forages needs to be calculated to provide the most economic return for producers. Livestock feeds are classified as either roughage or a forage—roughages have a crude fiber content of more than 18 percent, and concentrates have crude fiber content of less than 18 percent. Roughages can be either legume or nonlegume crops. A legume crop such as soybeans uses nitrogen from the air and is higher in protein content than nonlegume crops such as grass hay. Concentrates are made from either energy feeds (such as corn and oats) or from protein supplements provided by animal or vegetable sources.

A ration (the amount of feed an animal is given during a 24-hour period) must include the right balance of roughages and concentrates. A ration is only balanced if it provides an animal with all the nutrients it needs for growth, daily rate of gain, or production. A properly balanced ration should have an acceptable taste for the livestock as well as be cost effective for the producer. Any feed additives must be managed to not give higher amounts of certain substances than may be healthy for the animal. When balancing a ration for livestock, the needs for protein, energy, minerals, and vitamins must all be considered. There are five common steps that should be followed as a general rule of thumb when balancing a ration:

1. Identify the kind of animal to be fed.
2. Determine the animal's needs.
3. Select feeds and determine the composition of the feed.
4. Calculate the amount of each feed to use.
5. Check the ration against the animal's needs to make sure it is balanced.

Modern livestock operations use computer program software, and apps developed to assist in balancing feed rations for livestock. When balancing a ration by hand, the Pearson Square is a valuable tool. In this chapter, a dry-matter calculation problem will be presented as well as a sample ration that must be balanced.

Name_______________________________________ Date____________________

Exercise 8-1 Dry-Matter Calculation

Dry matter (DM) is the moisture-free content of a feed sample. The moisture dilutes the concentration of nutrients but does not have a major influence on intake. It is important to always balance and evaluate rations on a dry-matter basis. The calculation for converting nutrient values into a percentage of dry-matter value is as follows: The DM composition can be found by dividing the as-is value by the moisture content. (DM = As-is/Moisture content) For example, a feed with a 16.8 percent crude protein (CP) as-is ÷ 83.2 moisture content = 20.19% CP on a DM basis.

Directions: For each of the following feed analysis reports, convert each value into a dry-matter basis using the DM calculation.

CLIENT SAMPLE ID 1: Alfalfa Haylage

	-ANALYSIS-	
	AS-RECEIVED BASIS	DRY-MATTER BASIS
MOISTURE, %	67.6	0.0
DRY MATTER, %	32.4	100.0
CRUDE PROTEIN, %	5.9	______
HEAT DAM. PROTEIN, %	0.5	______
AVAILABLE PROTEIN, %	5.9	______
DIG. PROTEIN EST., %	3.8	______
ACID DET. FIBER, %	12.1	______
NEUT. DET. FIBER, %	17.6	______
TDN EST., %	19.1	______
ENE EST., THERMS/CWT	16.1	______
NE/LACT, MCAL/LB	0.19	______
NE/MAINT, MCAL/LB	0.19	______
NE/GAIN, MCAL/LB	0.10	______
PHOSPHORUS (P), %	0.11	______
CALCIUM (Ca), %	0.42	______
POTASSIUM (K), %	0.90	______
MAGNESIUM (Mg), %	0.11	______

Name____ Date__________________

CLIENT SAMPLE ID 2: Brown Midrib Sorghum Silage

	-ANALYSIS-	
	AS-RECEIVED BASIS	DRY-MATTER BASIS
MOISTURE, %	75.9	0.0
DRY MATTER, %	24.1	100.0
CRUDE PROTEIN, %	2.6	______
HEAT DAM. PROTEIN, %	0.3	______
AVAILABLE PROTEIN, %	2.5	______
DIG. PROTEIN EST., %	1.7	______
ACID DET. FIBER, %	9.4	______
NEUT. DET. FIBER, %	14.0	______
TDN EST., %	15.5	______
ENE EST., THERMS/CWT	13.1	______
NE/LACT, MCAL/LB	0.16	______
NE/MAINT, MCAL/LB	0.16	______
NE/GAIN, MCAL/LB	0.09	______
PHOSPHORUS (P), %	0.06	______
CALCIUM (Ca), %	0.12	______
POTASSIUM (K), %	0.49	______
MAGNESIUM (Mg), %	0.04	______

Name______________________________ Date______________

CLIENT SAMPLE ID 3: Baled Corn Stalks

	-ANALYSIS-	
	AS-RECEIVED BASIS	DRY-MATTER BASIS
MOISTURE, %	14.7	0.0
DRY MATTER, %	85.3	100.0
CRUDE PROTEIN, %	5.3	_____
HEAT DAM. PROTEIN, %	0.9	_____
AVAILABLE PROTEIN, %	4.7	_____
DIG. PROTEIN EST., %	3.6	_____
ACID DET. FIBER, %	41.6	_____
NEUT. DET. FIBER, %	59.4	_____
TDN EST., %	40.9	_____
ENE EST., THERMS/CWT	33.8	_____
NE/LACT, MCAL/LB	0.41	_____
NE/MAINT, MCAL/LB	0.35	_____
NE/GAIN, MCAL/LB	0.14	_____
PHOSPHORUS (P), %	0.18	_____
CALCIUM (Ca), %	0.35	_____
POTASSIUM (K), %	1.68	_____
MAGNESIUM (Mg), %	0.13	_____

Name______________________________ Date________________

Exercise 8-2 Balancing a Ration Using Equations

The Pearson Square (Figure 8-1) ration formulation procedure is designed to balance simple rations. For the square to work, follow specific directions for its use. The nutrient contents of ingredients and nutrient requirements must be expressed on the same basis (that is, dry-matter or "as-fed").

Several numbers are in and around the square, and one of the important numbers is the number that appears in the middle of the square. This number represents the nutritional requirement of an animal for a specific nutrient. It may be crude protein or TDN, amino acids, minerals, or vitamins.

In order to make the square work consistently, there are three very important considerations:

1. The value in the middle of the square must be between the two values that are used on the left side of the square. For example, the 14 percent crude protein requirement has to be an intermediate value between the soybean meal that has 45 percent crude protein or the corn that has 10 percent crude protein.
2. Disregard any negative numbers that are generated on the right side of the square. Be concerned only with the numerical differences between the nutrient requirement and the ingredient nutrient values.
3. Subtract the nutrient value from the nutritional requirement on the diagonal and arrive at a numerical value entitled parts. By summing those parts and dividing by the total, you can determine the percentage of the ration that each ingredient should represent to provide a specific nutrient level. Always subtract on the diagonal within the square to determine parts. Always double-check calculations to make sure that you do not have any mathematical errors. It also is very important to work on a uniform basis. Use a 100 percent dry-matter basis for nutrient composition of ingredients and requirements and then convert to an as-fed basis after the formulation is calculated.

   Corn represents (31.0/35.00) × 100 of the ration, or 88.57 percent.

   Soybean meal represents (4.0/35.00) × 100 of the ration, or 11.43 percent.

   Check of the calculation:

88.57 lb corn × 10.0% CP	=	8.86
11.43 lb SBM × 45.0% CP	=	5.14
100.00 lb mixture contains	=	14.00 lb CP, or 14 percent

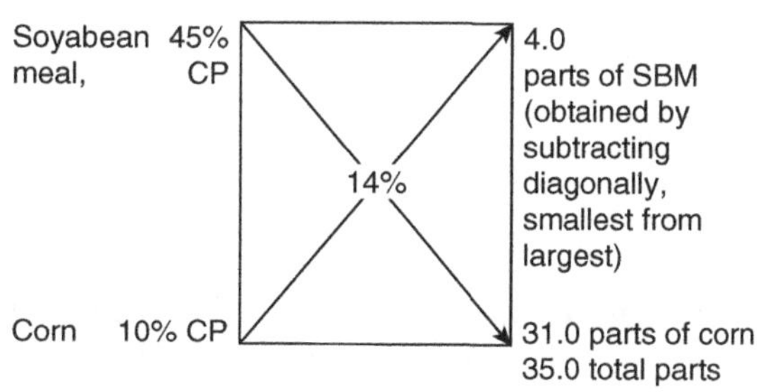

Figure 8-1 Pearson Square.

Name____ Date____________________

Directions: For the next two feeding scenarios, calculate the amount of each feed needed to create a balanced ration.

Scenario 1

You are feeding 20-pound baby pigs. Pigs this size need an 18 percent protein ration. You have shelled corn, which has 9 percent protein in it, but this is not enough protein. So, you decide to mix soybean oil meal with the shelled corn. The soybean oil meal has 40 percent protein in it. Mix the shelled corn and soybean oil meal together to come up with a 12 percent protein ration.

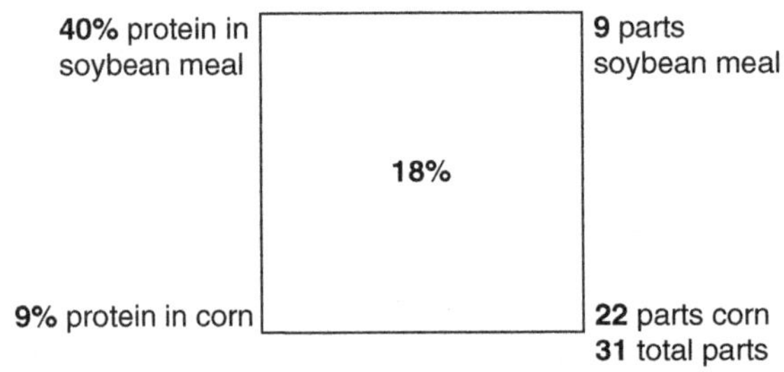

1. If you are going to mix 100 pounds of feed, how much corn do you need?

____

____

2. How much soybean meal do you need for 100 pounds of feed?

____

____

Scenario 2

Your pigs are now 50 pounds and you need to change the protein content in their ration. Now the ration must have 16 percent protein. You are going to mix corn, which has 9 percent protein, and soybean oil meal, which has 40 percent protein in it. What will you do?

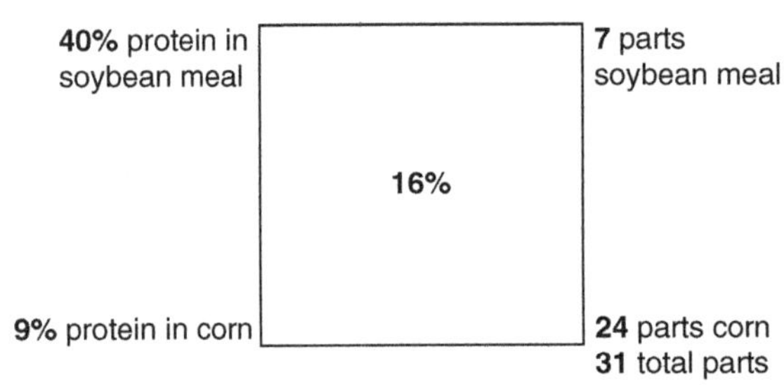

Name____ Date____________________

1. If you are going to mix 100 pounds of feed, how much corn do you need?

____

____

2. How much soybean oil meal do you need?

____

____

3. Does this feed ration have more corn or more soybean meal than in the previous ration? Why?

____

____

Name____ Date__________________

Chapter 8 Matching Activity

Directions: Match the term to the corresponding definition or description.

Term

____ 1. legume

____ 2. protein supplement

____ 3. energy feed

____ 4. ration

____ 5. balanced ration

____ 6. micronutrients

____ 7. digestible energy

____ 8. metabolizable energy

____ 9. dry-matter basis

____ 10. as-fed (air dry) basis

____ 11. animal protein

____ 12. vegetable protein

____ 13. tankage

____ 14. commercial protein supplement

____ 15. marbling

____ 16. gross energy

____ 17. total digestible nutrients (TDN)

____ 18. free choice

____ 19. palatable

____ 20. bomb calorimeter

Definitions

a. protein supplements that come from animals and animal by-products and contain more than 47 percent crude protein

b. livestock feeds with less than 20 percent crude protein

c. protein supplements that come from plants

d. ingredients used in small quantities in the diet

e. animal tissues and bones from animal slaughterhouses and rendering plants that are cooked, dried, and ground

f. the gross energy of a feed minus the energy remaining in the feces of the animal after the feed is digested

g. commercial protein supplement

h. plants that have nodules (small swellings or lumps) on their roots that contain nitrogen-fixing bacteria

i. intramuscular fat, especially in beef, that adds flavor, taste, and tenderness to the meat

j. the mass of a material when the moisture content is removed

k. the total amount of heat released by burning in the bomb calorimeter

l. it gives a measure of the total energy value of a feed when it is fed to an animal

m. a supplement that is available at all times to the animal

n. livestock feeds that contain 20 percent or more protein

o. data that are calculated on the basis of the average amount of moisture found in the feed as it is used on the farm

p. a lab device used to burn (oxidize) feed to obtain gross energy

q. a ration that has all the nutrients the animal needs in the right proportions and amounts

r. the gross energy in the feed eaten minus the energy found in the feces, the energy in the gaseous products of digestion, and the energy in the urine

s. the amount of feed given to an animal to meet its needs during a 24-hour period

t. a ration must taste good and meet the nutritional requirements of the animal

Name_________________________________ Date_______________

Chapter 8 Lab Questions

1. What is the difference between roughages and concentrates?

2. Describe the characteristics of a well-balanced ration.

3. What are five common steps to use when balancing a ration?

4. How does the Pearson Square work?

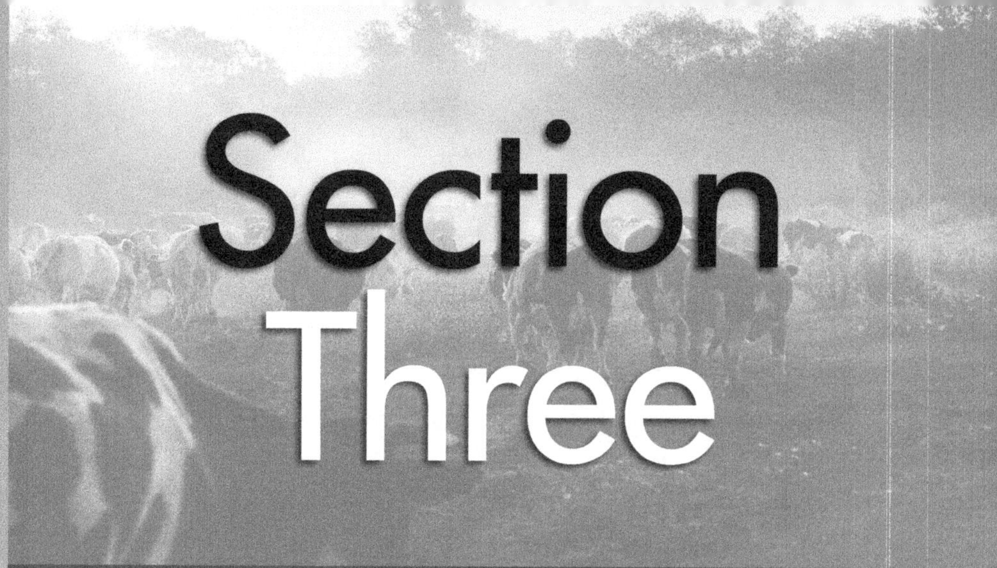

Section Three

Animal Breeding

Feel good studio/Shutterstock.com

Feel good studio/Shutterstock.com

Chapter 9

Genetics of Animal Breeding

Introduction

Modern livestock have improved in quality from livestock of the past due to advancements in, and increased understanding of, animal genetics. Livestock producers are able to use genetic technology to select for desirable traits among their herds, and over time, they can change the genetic makeup of the entire herd. Genetics is the study of heredity, or the way in which traits of parents are passed on to offspring. This concept was founded by an Austrian monk named Gregor Mendel. From 1857 to 1865, Mendel conducted many experiments with garden peas in order to prove that certain characteristics were passed on from the parent to the offspring.

The cells of an animal's body undergo cell division to allow the body to continually grow. Within the nucleus of each cell, DNA is contained in the form of chromosome pairs. One chromosome of the pair comes from the mother and the other comes from the father. The amount of difference between parents and offspring is caused by genetics and environmental factors. Heritability estimates are used to show how much of the difference in some traits may come from genetics.

Genes are complex molecules found on the chromosomes; genes have the ability to affect the dominant traits displayed in the animal. Some genes are dominant and others are recessive. Dominant genes overpower the effect of recessive genes. Some genes are neither dominant nor recessive; as a result, a mixture of the two gene effects is displayed.

An understanding of genetic probability is important for livestock producers to influence the quality of their herd genetics over time. Desirable traits improve animal efficiency and economic return from products. In this chapter, the Punnett square will be mastered through several genetic scenarios that need to be calculated to determine which traits will be inherited by offspring. An activity identifying common recessive and dominant traits will also demonstrate the practical use of animal genetics.

Name________________________________ Date________________

Exercise 9-1 Punnett Square

Directions: Answer the following questions based on the information provided.

Use the following information for questions 1–3:

In dogs, the gene for fur color has two alleles. The dominant allele (F) codes for grey fur, and the recessive allele (f) codes for black fur.

1. The female dog is heterozygous. The male dog is homozygous recessive. Figure out the phenotypes and genotypes of their possible puppies by using a Punnett square.

Genotypes	**Phenotypes**
FF: __________	Black fur: __________
Ff: __________	Grey fur: __________
ff: __________	

2. The female dog has black fur. The male dog has black fur. Figure out the phenotypes and genotypes of their possible puppies by using a Punnett square.

Genotypes	**Phenotypes**
FF: __________	Black fur: __________
Ff: __________	Grey fur: __________
ff: __________	

3. The female dog is heterozygous. The male dog is heterozygous. Figure out the phenotypes and genotypes of their possible puppies by using a Punnett square.

Genotypes	**Phenotypes**
FF: __________	Black fur: __________
Ff: __________	Grey fur: __________
ff: __________	

Use the following information for questions 4–6:

In fruit flies, red eyes are dominant (E), and white eyes are recessive (e).

4. If the female fly has white eyes and the male fly has homozygous dominant red eyes, what are the possible phenotypes and genotypes of their offspring?

Genotypes	**Phenotypes**
EE: __________	Red eyes: __________
Ee: __________	White eyes: __________
ee: __________	

Name______________________________________ Date__________________

5. If the female fly has EE and the male fly has EE, what are the possible phenotypes and genotypes of their offspring?

Genotypes	**Phenotypes**
EE: ____________________	Red eyes: ____________________
Ee: ____________________	White eyes: ____________________
ee: ____________________	

6. If both flies are heterozygous, then what are the possible phenotypes and genotypes of their offspring?

Genotypes	**Phenotypes**
EE: ____________________	Red eyes: ____________________
Ee: ____________________	White eyes: ____________________
ee: ____________________	

Use the following for questions 7–9:

In dogs, there is a hereditary deafness caused by a recessive gene, "d." A kennel owner has a male dog (Gilbert) that she wants to use for breeding purposes if possible. The dog can hear.

7. What are the two possible genotypes of Gilbert?

____

____

8. If the dog's genotype is Dd, the owner does not wish to use him for breeding so that the deafness gene will not be passed on. This can be tested by breeding the dog to a deaf female (dd). Draw two Punnett squares to illustrate these two possible crosses.

____

____

9. In each case, what percentage/how many of the offspring would be expected to be hearing? Deaf? How could you tell the genotype of this male dog? Also, using Punnett square(s), show how two hearing dogs could produce deaf offspring.

____

____

A Guide to Using the Punnett Square

The value of studying genetics is understanding how to predict the likelihood of an animal inheriting particular traits. This can help to develop plants and animals that have more desirable qualities.

One of the easiest ways to calculate the probability of inheriting a specific trait was invented in the early twentieth century by a geneticist named Reginald Punnett. He developed a technique to predicting genetic traits that we now call a Punnett square. This is a simple method of discovering all of the potential combinations of genotypes that can occur in offspring, given the genotypes of their parents. It also shows us the odds of each of the offspring genotypes occurring.

Name____ Date____________________

Setting up and using a Punnett square is quite simple once you understand how it works. You begin by drawing a grid of perpendicular lines:

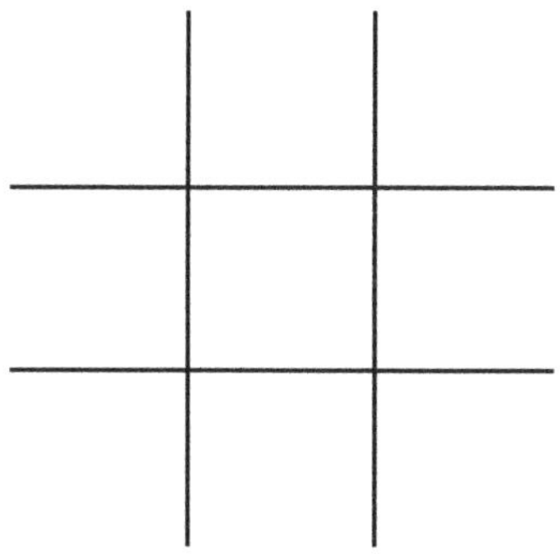

Next, you put the genotype of one parent across the top and that of the other parent down the left side. For example, if parent pea plant genotypes were YY and GG, respectively, the setup would be:

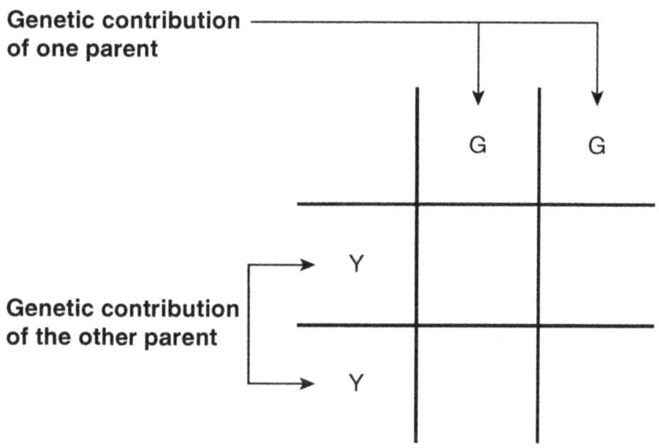

Note that only one letter goes in each box for the parents. Next, all you have to do is fill in the boxes by copying the row and column-head letters across or down into the empty squares. This gives us the predicted frequency of the potential genotypes among the offspring each time reproduction occurs.

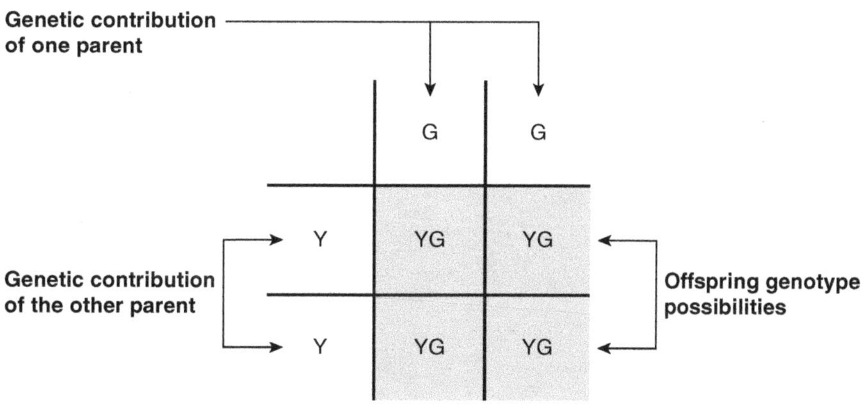

Name____ Date____________________

In this example, 100 percent of the offspring will likely be heterozygous (YG). Since the Y (yellow) allele is dominant over the G (green) allele for pea plants, 100 percent of the YG offspring will have a yellow phenotype, as Mendel observed in his breeding experiments.

In the following example, if the parent plants both have heterozygous (YG) genotypes, there will be 25 percent YY, 50 percent YG, and 25 percent GG offspring on average. These percentages are determined based on the fact that each of the four offspring boxes in a Punnett square is 25 percent (1 out of 4). As to phenotypes, 75 percent will be Y and only 25 percent will be G. These will be the odds every time a new offspring is conceived by parents with YG genotypes.

	Y	G
Y	YY	YG
G	YG	GG

An offspring's genotype is the result of the combination of genes in the sex cells or cells that came together in its conception. One sex cell came from each parent. Sex cells normally only have one copy of the gene for each trait. Each of the two Punnett square boxes in which the parent genes for a trait are placed (across the top or on the left side) actually represents one of the two possible genotypes for a parent sex cell. Which of the two parental copies of a gene is inherited depends on which sex cell is inherited—it is a matter of chance. Placing each of the two copies in its own box has the effect of giving it a 50 percent chance of being inherited.

Source: Dennis O'Neil.

Name______________________________ Date_______________

Exercise 9-2 Dominant vs. Recessive Allele Activity

A dominant gene in a heterozygous pair hides the effect of its allele. The allele that is hidden is called a recessive gene. The polled (not having horns) condition in cattle is the result of a dominant gene and is said to be a dominant trait.

Directions: Using the word bank provided, place each trait in the appropriate column, either dominant or recessive.

Word Bank:

- Black coat in horses
- Color in animals
- Red coat in cattle
- Normal size in cattle
- Black coat in cattle
- Barred feather pattern in chickens (also sex-linked)
- Nonbarred feather pattern in chickens
- Rose comb in chickens
- Pea comb in chickens
- Colored face in cattle
- White face in cattle
- Albinism in animals
- "Snorter" dwarfism in cattle
- Single comb in chickens
- Brown coat in horses

Dominant	Recessive

Name____ Date____________________

Chapter 9 Matching Activity

Directions: Match the term to the corresponding definition or description.

Term

____ 1. genetics
____ 2. genotype
____ 3. phenotype
____ 4. additive gene effect
____ 5. nonadditive gene effect
____ 6. heritability
____ 7. heritability estimate
____ 8. breeding value
____ 9. gene
____ 10. allele
____ 11. heterosis
____ 12. chromosome
____ 13. mitosis
____ 14. meiosis
____ 15. gamete
____ 16. ribonucleic acid
____ 17. codon
____ 18. mutation
____ 19. homozygous gene pair
____ 20. heterozygous gene pair

Definitions

a. the proportion of the total variation that is due to the additive gene effects
b. sometimes called hybrid vigor, when combinations of gene pairs of the offspring will be better than either of their parents
c. when many different genes are involved in the expression of a trait
d. contains the genes
e. an estimate that helps the producer to make faster genetic improvement in livestock
f. the division of body cells; the chromosomes pairs are duplicated in each daughter cell, so that they are exactly alike
g. each gene in a gene pair; codes for a different expression of the same trait
h. is a cell division that created gametes and the number of chromosomes in the gametes is reduced by half
i. expresses the likelihood of a trait being passed from one parent to offspring
j. the reproductive cells of an organism; known as the sperm and egg cells
k. the study of heredity
l. regulates protein synthesis in animals
m. these traits are readily observable and are controlled by only one or a few gene pairs
n. the basic unit of the genetic code
o. a unit of heredity that is transferred from a parent to offspring
p. one that carries two genes for a trait
q. the combination of genes that an individual possesses
r. is a change in the DNA sequence of an organism
s. the physical appearance of an animal
t. one that carries two different genes (called alleles) that affect a trait

Name________________________________ Date______________

Chapter 9 Lab Questions

1. Who was Gregor Mendel? What was his contribution to the study of genetics?

2. Who was Reginald Punnett? What was his contribution to the study of genetics?

3. How does the Punnett square work? Explain.

4. What are hereditability estimates used for?

Feel good studio/Shutterstock.com

Chapter 10

Animal Reproduction

Introduction

Through mating, female and male animals begin the intricate process of reproduction. The male gamete, called the sperm, is deposited in the female reproductive tract, where it moves to the oviduct and fertilizes the egg cell (female gamete). In livestock, all the females go through a period during which they will accept the male for mating; this is called the estrus cycle. During estrus, otherwise known as the heat period, hormones cause the egg cell to be released for fertilization.

If the egg cell is fertilized, it moves to the uterus, where it grows into a fetus; pregnancy is maintained by hormones and body functions. When the fetus has developed for a full term and is ready for birth, the uterine muscles contract, forcing the fetus through the birth canal. Reproduction in poultry is similar in some ways to reproduction in mammals. The main difference is that the embryo develops in the eggshell outside of the mother's body. Eggs must be kept at the proper temperature and humidity to hatch. This is usually done in an incubator.

On many livestock farms, animals are raised in separate groups depending on their reproductive development. If animals are placed in the wrong reproductive group, many negative outcomes are possible: (1) animals may be stressed and lower in daily weight gain or production, (2) animals that should not be bred due to health problems are at risk, (3) young animals can die from premature pregnancies if placed in the wrong group, and (4) young males can be stressed and injured by aggressive older livestock in the same group.

When managing livestock reproduction, knowledge of anatomy is very important, as well as understanding the normal functions of the reproductive system during pregnancy and birth. If an animal has difficult labor, a manager must be able to recognize the problem and assist with the birth without harming the cow or the calf. In this chapter, you will practice grouping animals at different reproductive stages of life, as well as identifying reproductive anatomy parts.

Name______________________________ Date______________

Exercise 10-1 Reproductive Planning for Livestock

A beef cattle operation must manage the reproductive cycles and interactions between male and female animals on the farm. When raising cattle, knowledge of the reproductive development is a key factor to select animals for breeding purposes. A producer must be able to identify the stage of reproductive development that animals are at and separate livestock based on the breeding strategy. If a cow is "open," it means it is not pregnant.

The following list shows a mixed group of cattle at different stages of development. Each animal will need to be separated into the appropriate group for development. Study the herd records, and identify which animals belong in each group of reproductive development. Place the proper animal identification number in the group it should be integrated into.

Mixed Cattle Group: Herd List

Identification Number	Gender/Age	Animal Status
1012	Male/24 months	Castrated/weighs 1,200 lb
9000	Female/52 months	Open/2-month-old calf at side
2611	Female/7 months	Open/650 lb
1029	Male/10 months	Uncastrated/900 lb
8826	Female/46 months	Open/1-week-old calf at side
2020	Female/120 months	Open last two breeding cycles
1893	Female/12 months	Open/weighs 950 lb
1982	Male/18 months	Castrated/weighs 1,100 lb
1729	Female/24 months	8 months pregnant
2001	Female/36 months	5 months pregnant
4233	Female/9 months	Open/weighs 800 lb
1482	Female/36 months	Open/aborted last year's calf
1673	Male/48 months	Uncastrated herd sire
8675	Male/7 months	Castrated/weighs 700 lb
9212	Female/14 months	Open/weighs 950 lb
3420	Male/7 months	Uncastrated/750 lb
4200	Female/95 months	3 months pregnant

Name____ Date____________________

Directions: Place the individual cattle into the appropriate groups. List the identification number and provide a reason that animal was placed in this group.

Bulls that are old enough for breeding and steers that are over 800 pounds:

____

____

Cattle that are too old for reproduction:

____

____

Cattle that are too young or small for breeding:

____

____

Cattle that need to be bred and are the correct age and weight:

____

____

Cattle that are of age and size, but should not be bred currently:

____

____

Name______________________________ Date______________

Exercise 10-2 Reproductive Anatomy Activity

Special Instructions: Reproductive tracts can be obtained from a local slaughterhouse or possibly a livestock producer. Class members performing the dissection should use caution because the scalpel can become slippery during the dissection. All participants should wear gloves as protection against any disease organisms that may be present. Using the following anatomy diagrams (**Figures 10-1** through **10-5**), identify the parts of each sample reproductive tract.

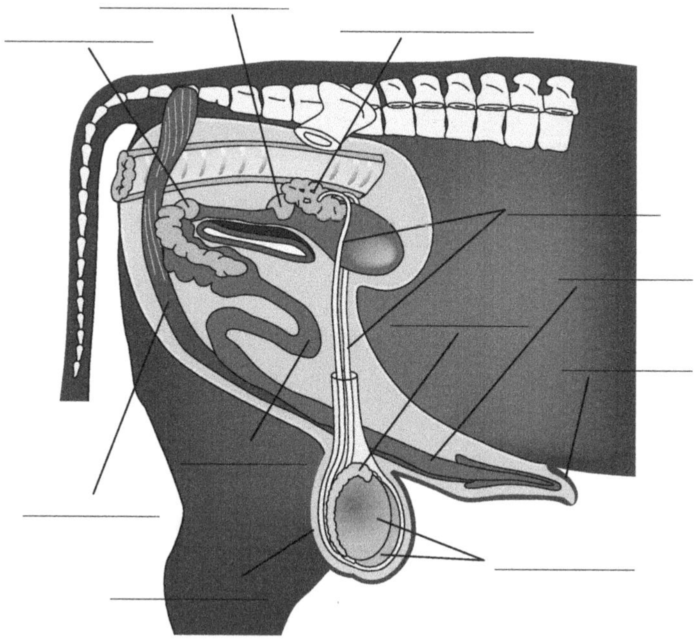

Figure 10-1 Reproductive organs of a bull.

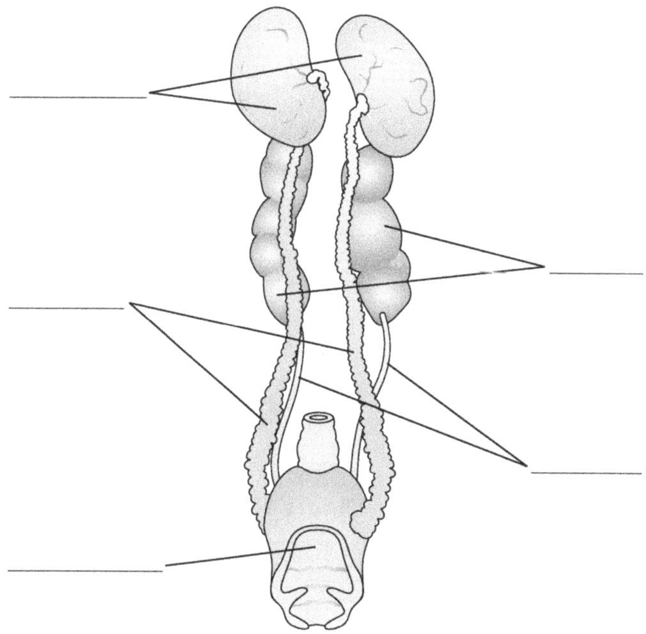

Figure 10-2 Reproductive system of the male chicken.

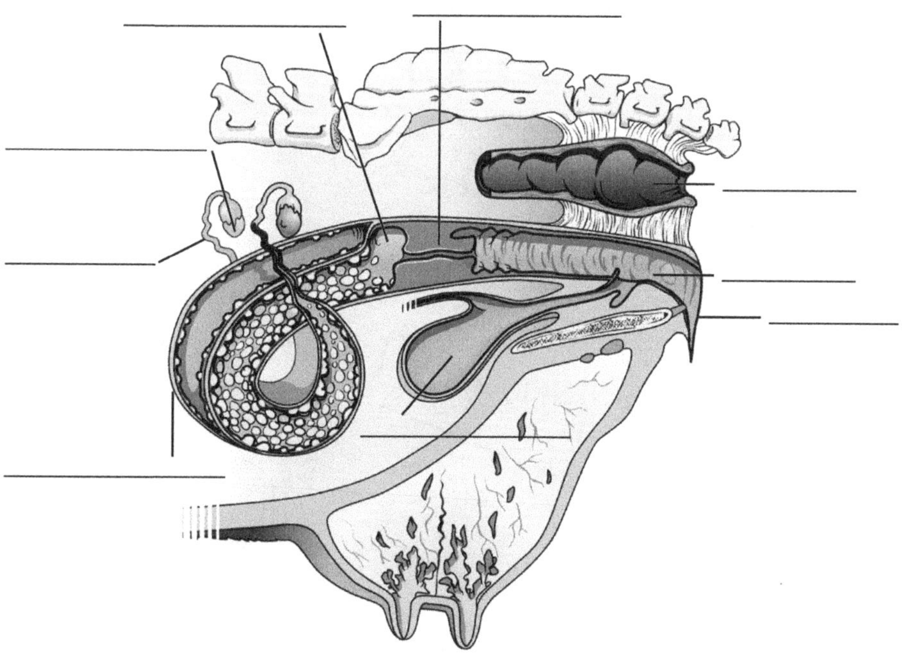

Figure 10-3 Female reproductive system of a cow, side view.

Name____________________ Date__________

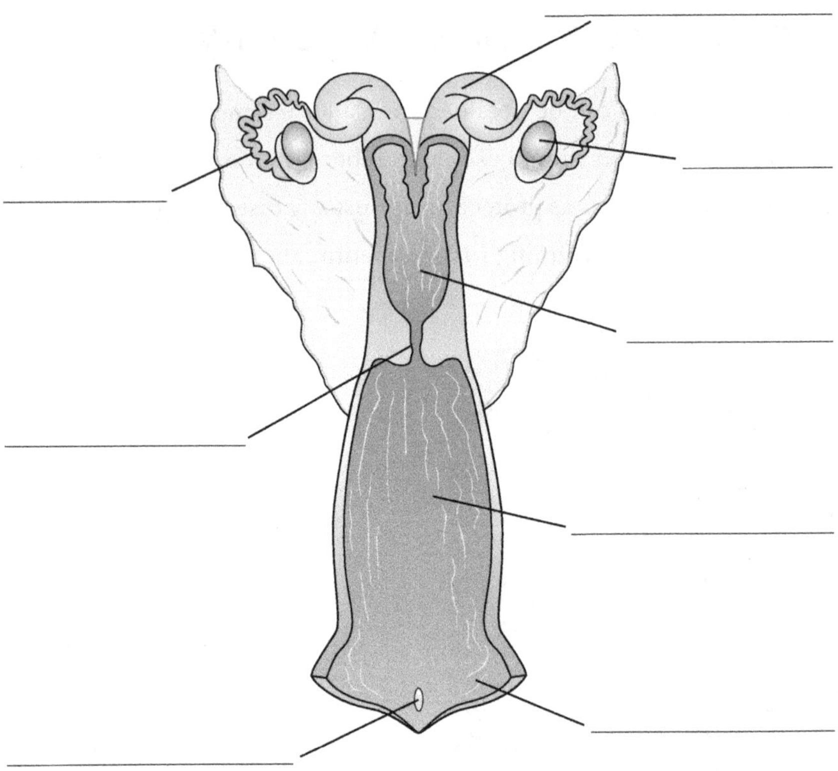

Figure 10-4 Reproductive system of a female cow, dorsal view.

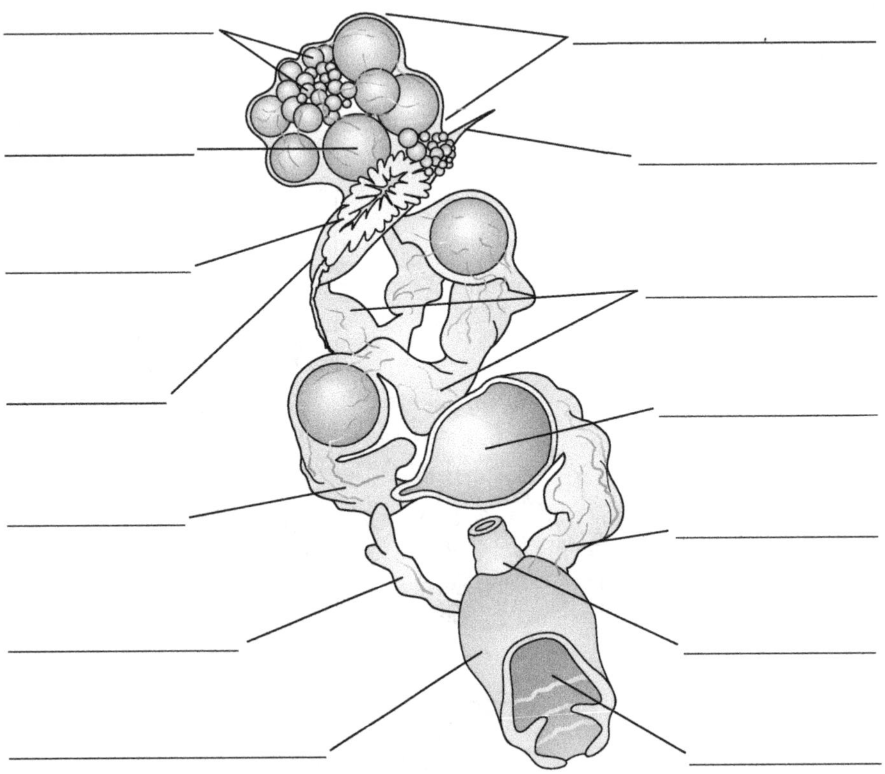

Figure 10-5 Reproductive system of the female chicken.

Name____ Date____________________

Part 1: Female Anatomy

Materials needed:

Reproductive tract from cow, ewe, hen, or sow

Surgical gloves for each class member

Scalpels

Dissecting pins and labels

Plastic aprons for each class member

Procedure:

1. Lay tracts from different species side by side so that comparisons can be made.
2. Prior to dissecting the tracts, identify the ovaries, infundibulum, oviducts, uterine horns, uterine body, cervix, vagina, and vulva. Note differences between species. Label parts with dissecting pins.
3. Identify any structures on the ovaries.
4. Dissect the uterus of the cow and/or ewe and note the septum that divides what, from the outside, appears to be the uterine body.
5. Dissect the cervix of each species. Identify rings or folds in the cervix. Note differences between species.
6. Dissect the vagina of each species. Locate the urethral opening in the floor of the vagina. Locate the external os at the junction of the cervix and vagina. Compare the external os of the cow and/or ewe to the external os of the sow. Identify the clitoris, which is located at the floor of the vagina where the vagina joins the vulva. Label parts with dissecting pins.
7. Locate semen deposition sites in each species for natural service and artificial insemination. Label with dissecting pins.

Part 2: Male Anatomy

Materials needed:

Reproductive tracts from bull, ram, rooster, or boar

Surgical gloves for each class member

Scalpels

Dissecting pins and labels

Procedure:

1. Lay tracts from different species side by side so that comparisons can be made.
2. Identify the scrotum, testicle, epididymis, vas deferens, urethra, Cowper's gland, prostate, seminal vesicles, and penis. Locate the sigmoid flexure of the penis. Label all parts with dissecting pins.
3. Dissect one testicle and locate the ducts of the rete testes. Label with dissecting pins.

Name____ Date____________________

Chapter 10 Matching Activity

Terms

____ 1. parturition
____ 2. gestation
____ 3. estrogen
____ 4. sexual reproduction
____ 5. asexual reproduction
____ 6. testosterone
____ 7. corpus luteum
____ 8. uterus
____ 9. alimentary canal
____ 10. dystocia
____ 11. colostrum milk
____ 12. epididymis
____ 13. cyst
____ 14. infundibulum
____ 15. Cowper's gland
____ 16. urethra
____ 17. cryptorchidism
____ 18. ovulation
____ 19. afterbirth
____ 20. estrus

Definitions

a. the act of giving birth; the final step in reproduction
b. is a heat cycle/period that occurs in female mammals and is the time during which the female will accept the male for breeding
c. period of time during which the animal is pregnant
d. a Y-shaped structure consisting of the body, two uterine horns, and the cervix
e. is a tube found in both male and female mammals that urine moves through
f. an inherited trait if the testicles of a male animal are held in the body cavity, the animal is sterile, or the animal cannot produce live sperm
g. the follicle ruptures and releases the ovum
h. a long, coiled tube that is connected to each testicle
i. female sex hormones
j. the placenta and other membranes that were not expelled when the fetus was born
k. the union of the male and female gametes
l. male hormones
m. is a food-carrying passage that begins at the mouth and ends at the vent in poultry
n. difficulty in birthing
o. produces a fluid that moves down the urethra ahead of seminal fluid
p. a swelling containing fluid or a semi-solid substance
q. the first milk given to a newborn animal during the first 12 to 24 hours after birth
r. the funnel-shaped end of the oviduct that catches the ovum
s. reproduction that does not involve the gametes
t. ovaries from females that produce progesterone

Name______________________________ Date______________

Chapter 10 Lab Questions

1. Explain estrus and how it functions.

2. Describe the problems that can arise during livestock parturition.

3. Why is it important that animals be grouped according to reproductive status?

Feel good studio/Shutterstock.com

Chapter 11

Biotechnology in Livestock Production

Introduction

Since early human civilization, humans have been using some forms of biotechnology to change living plants and animals for commercial use. Activities as simple as livestock breeding, crop improvement, and the production of food products are basic forms of biotechnology. As microbiology and technology increased in the 1970s, laboratory techniques were developed that allowed researchers to identify and manipulate the DNA found in cells of living organisms. This DNA contains information about an organism's genetic makeup. As this technology has improved over

Glynnis Jones/Shutterstock.com

Figure 11-1 Sign against the use of genetically modified products in agriculture.

the past 50 years, the ability to influence many characteristics of living organisms has increased. The science of altering genetic and reproductive processes in animals and plants is called agricultural biotechnology.

Most of the work being done in research and biotechnology is currently focused on genetic engineering and embryo transfer. Genetic engineering involves taking a tiny bit of DNA containing the desired gene from one organism and splicing it onto the DNA strand in another organism. The recipient organism takes on the characteristic controlled by the transferred gene. There are many forms of genetically engineered techniques and substances that can be administered to animals to affect their performance or characteristics. Many other areas of research are currently being conducted with genetic engineering in animal and plant science. One example includes research to identify genetic markers for cattle and swine that make it easier to breed animals for greater resistance to disease and parasites.

Another major technique from biotechnology is embryo transfer. This has become an established technology in cattle production. In embryo transfer, the embryos from an animal with desirable genetic characteristics are "flushed" out of the animal; the embryo is fertilized and then implanted in a surrogate animal to carry out the pregnancy. The use of embryo transfer permits the production of many more offspring from genetically desirable animals.

Genetically modified organisms (GMOs) are very beneficial in terms of increased production of agricultural products in a world with a growing population, but the long-term impacts of this technology cause concern in consumers and activists. In this chapter, debates around the use of biotechnology are explored.

DC Studio/Shutterstock.com

Figure 11-2 Scientist researching new ways to utilize biotechnology.

Name_______________________________________ Date____________________

Exercise 11-1 Using rbST

An example of genetic engineering used in dairy cattle was the development of recombinant bovine somatotropin (rbST), the manufactured form of bovine somatotropin (bST), a natural occurring growth hormone that causes energy derived from feed to be used for milk production rather than for weight gain. Data showed that the use of rbST increased milk production in dairy cattle by 10 to 15 percent. The FDA approved the use of rbST in 1993 after almost 10 years of testing determined that milk and meat from cows treated with rbST was safe for human consumption. However, consumer resistance to milk from rbST-treated cows has resulted in rbST rarely being used today. Since bovine somatotropin is a natural hormone and trace amounts of it can be found in all milk, the FDA ruled that milk from cows not treated with rbST cannot be labeled "bST free." Sellers, however, are permitted to advertise products as "rbST free."

Directions: Using additional information in the textbook and online sources, make a list of positive and negative effects of using rbST in dairy cattle. Once finished, come to a conclusion that includes your own opinion about the use of rbST. Explain your reasoning.

Name______________________________ Date______________

Exercise 11-2 Genetically Modified Organisms Debate Activity

Pre-debate Exercise

Directions: First research and read two articles about GMOs using online resources. One article should be supportive of using GMO technology in our food system and the other should be against the use of GMOs. After reading and researching, complete the following questions and tasks using facts and arguments found in the readings.

1. What are genetically modified crops?

2. Make a list of positive effects of genetically modified crops.

3. Make a list of negative effects or risks associated with genetically modified crops.

4. How do genetically modified crops affect the hunger crisis in the world?

5. Is there a way to supply food to the world without using GMO products?

Name______________________________________ Date__________________

Four-Corner Debate Activity

When defending an opinion during a debate, one must study both sides of an argument in order to demonstrate a true knowledge of the topic and establish credibility for an opinion.

Directions: In the classroom, each of the four corners will have a sign posted designating an opinion. The signs will read (1) agree, (2) strongly agree, (3) disagree, and (4) strongly disagree. When the instructor reads a controversial statement about GMOs, students must decide which sign best represents their opinion and go stand next to the sign. The instructor will then ask each student to defend his or her opinion with a well-educated explanation.

Name____ Date____________________

Chapter 11 Matching Activity

Directions: Match the term to the corresponding definition or description.

Term	Definition
____ 1. biotechnology	**a.** proteins that affect the utilization of energy in the body
____ 2. transgenic	**b.** the science of altering genetic and reproductive processes in animals and plants
____ 3. clone	**c.** by utilizing this technology, it is possible to produce many more offspring from desirable animals than would be possible if the animals had to carry each embryo to term before producing another one
____ 4. genetic engineering	**d.** the process in which the gene that controls bST production is spliced into the DNA of bacteria, which is then injected into the cow
____ 5. somatotropins	**e.** when cells or organisms are genetically identical to each other
____ 6. bovine somatotropin	**f.** manipulating the DNA of living organisms at the molecular and cellular level to produce new commercial applications
____ 7. recombinant bovine somatotropin (rbST)	**g.** causes energy derived from feed to be used for milk production rather than for weight gain, but does not reduce the energy available for body maintenance
____ 8. superovulation	**h.** the process in which human genes are inserted into animal embryos
____ 9. agricultural biotechnology	**i.** the process of identifying and transferring a gene or genes for a specific trait from one organism to another
____ 10. CRISPR	**j.** used to edit genes

Name____ Date__________________

Chapter 11 Lab Questions

1. Describe ways biotechnology is being used in agriculture.

2. How does genetic engineering work? Describe the process.

3. How is cloning being used in the livestock industry?

4. How is embryo transfer used as a livestock management tool?

Name______________________________ Date______________

5. How is CRISPR being used in the animal industry?

____

____

____

____

____

____

Feel good studio/Shutterstock.com

Animal Breeding Systems

Introduction

Strategic livestock breeding can be beneficial in many ways by improving herd quality and performance factors, as well as by making the herd more economically efficient. The two basic systems of livestock breeding are straightbreeding and crossbreeding. The type of breeding system used depends on the farmer and the goals of the operation. Some livestock operations choose to maintain purebred animals for various reasons. Purebred animals can be eligible for registry in a breed association with documentation of the bloodline. Establishing a purebred herd takes intense management and investment in pure pedigree animals.

Purebred animals are not always the most advantageous for agricultural production. Sometimes, a crossbred animal will contain a combination of more desirable traits than a purebred could achieve. Many livestock producers choose to manage a crossbreeding system where two different breeds are combined to achieve the desired traits; the resulting offspring are called a hybrid. Superior traits that result from crossbreeding are called hybrid vigor, which is measured by the average superiority of the hybrid offspring over the average of the parents.

Research has shown that well-planned crossbreeding programs can increase total productivity in beef herds by 20 to 25 percent. In order to achieve this type of increase in productivity, the producer must follow a good performance selection program, good management, good nutrition, and good herd health practices to achieve the desired results. Most livestock producers use crossbreeding, except for most dairy producers and many breeds of horses.

Crossbreeding systems range from those that are very complex to those that are very simple, so each producer needs to assess which system works best for the current circumstances of the operation. Some of the common systems used are terminal sire crossed with F_1 females, rotating the herd bull every three or four years, two-breed rotation, three-breed rotation, four- and five-breed rotations, static terminal sire systems, rotational-terminal sire system, and composite breeding systems.

In this chapter, Exercise 12-1 provides examples of livestock operations that have a desired trait that they are trying to attain through breeding strategies. For each example, use the breeding techniques discussed in the textbook to suggest the proper breeding strategy to achieve the desired hybrid vigor.

Linebreeding

(A represents the male; B and C represent females)

1st mating: A X B A X C
1st generation: $\frac{1}{2}$ A and $\frac{1}{2}$ B $\frac{1}{2}$ A and $\frac{1}{2}$ C

2nd mating: $\frac{1}{2}$ A and $\frac{1}{2}$ B X $\frac{1}{2}$ A and $\frac{1}{2}$ C
2nd generation: $\frac{1}{2}$ A and $\frac{1}{4}$ B and $\frac{1}{4}$ C

The offspring in the second generation have received 50 percent of their genetic inheritance from the sire A because he appears twice in their pedigree. They have received only 25 percent of their genetic inheritance form each of the females B and C.

Grading Up

(A_1, A_2, A_3 represent purebred sires of a given breed; G represents a grade female)

1st mating: A_1 X G
1st generation: $\frac{1}{2}$ A_1 and $\frac{1}{2}$ G (50% purebred, 50% grade)

2nd mating: A_2 X $\frac{1}{2}$ A_1 and $\frac{1}{2}$ G
2nd generation: $\frac{1}{2}$ A_2 and $\frac{1}{4}$ A_1 and $\frac{1}{4}$ G (75% purebred, 25% grade)

3rd mating: A_3 X $\frac{1}{2}$ A_2 and $\frac{1}{4}$ A_1 and $\frac{1}{4}$ G
3rd generation: $\frac{1}{2}$ A_3 and $\frac{1}{4}$ A_2 and $\frac{1}{8}$ A_1 and $\frac{1}{8}$ G (87.5% purebred, 12.5% grade)

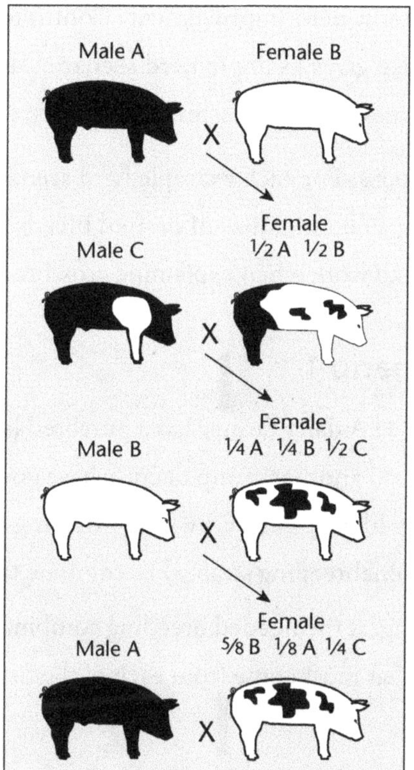

A, B, and C represent different breeds. Continue using males of three breeds in sequence. Using males of the three breeds in sequence is called rotating the males: hence, the inclusion of the term *rotation* in the name of this system.

Figure 12-1 Using Heterosis (Hybrid Vigor) to achieve desired traits.

Name____ Date____________________

Exercise 12-1 Improving Herd Quality Through Proper Breeding Strategy

In most livestock used for meat consumption, crossbreeding can result in a combination of many desirable traits and long-term herd improvement. Controlled crossbreeding takes a high level of management and record keeping. This exercise gives example herd scenarios and the desired combinations of breeds and traits that the farmer is trying to produce through crossbreeding strategies.

Directions: For each example herd scenario, show the combinations needed to use the existing herd and make the offspring a combination of desired breeds. All offspring should have the desired percentage of each breed. Make sure to show all work when explaining crossbreeding strategies.

Scenario 1

Herd 1: A dairy farmer has a purebred herd of Holstein cattle. Some of the cows produce a high level of milk every day (A), and another group of cows have good conformation and good feet and legs (B). The farmer has also just bought a high-pedigree bull (C) whose offspring have all proven to possess a good calving ease trait. The dairy farmer hopes to use a linebreeding strategy to combine these traits over four generations of offspring.

Suggest the needed breeding combinations and determine the percentage of inherited genetics in the fourth generation that must come from each of the first animals to be mated. Show all work.

Scenario 2

Herd 2: A beef cattle rancher has a herd of grade females that are not purebred. The rancher decides to use breeding strategy to improve their herd. The national price for beef from cows that are 90 percent Angus breed is greater than for beef from grade animals. The rancher plans to buy a new purebred Angus bull to sire one generation of offspring. A new bull will be introduced to the herd every generation for four generations. The letters A, B, C, and D represent the purebred Angus sires that will be used to help the beef operation grade up. G represents a grade female.

Name____ Date____________________

Demonstrate the change that will occur in the beef through four generations of using the grade-up strategy. Show in fractions the number of hereditary traits in the offspring that result from each of the original mated animals. Show all work.

Scenario 3

Herd 3: A hog farmer wants to improve the growth and performance of the herd and decides to use a three-breed rotational crossbreeding system. The farmer decides to use two sows from the herd to test this new breeding system. One of the females is a white Heritage breed, and the other female is a Yorkshire breed. The farmer will breed each of the sows with three different boars over three generations. The first generation will be sired by a Tamworth boar, the second generation will be sired by a Duroc boar, and the third generation will be sired by a Berkshire boar.

Demonstrate the combination of genetics that will take place in each generation for both the Heritage sow and the Yorkshire sow. Show all work.

Name____ Date____________________

Chapter 12 Matching Activity

Directions: Match the term to the corresponding definition or description.

Term	Definition
____ 1. straightbreeding	**a.** the animals being mated are very closely related and can be traced back to more than one common ancestor
____ 2. purebred	**b.** the mating of purebred sires to grade females
____ 3. inbreeding	**c.** mating animals from two different lines of breeding within a breed
____ 4. closebreeding	**d.** also known as heterosis, superior traits that result from crossbreeding
____ 5. linebreeding	**e.** the resulting offspring from crossbreeding
____ 6. outcrossing	**f.** an animal of a particular breed that has the characteristics of the breed to which it belongs
____ 7. linecrossing	**g.** the mating of two animals from different breeds
____ 8. grading up	**h.** the mating of related animals
____ 9. grade	**i.** the mating of animals that are more distantly related
____ 10. hybrid	**j.** the mating of animals of the same breed
____ 11. crossbreeding	**k.** superior traits that result from crossbreeding
____ 12. hybrid vigor	**l.** the mating of animals of different families within the same breed

Name______________________________ Date______________

Chapter 12 Lab Questions

1. What is the difference between straightbreeding and linebreeding?

2. Why is grading up used as a livestock breeding system?

3. What are the benefits that result from crossbreeding livestock?

4. What are some factors that affect breeding systems? Give examples.

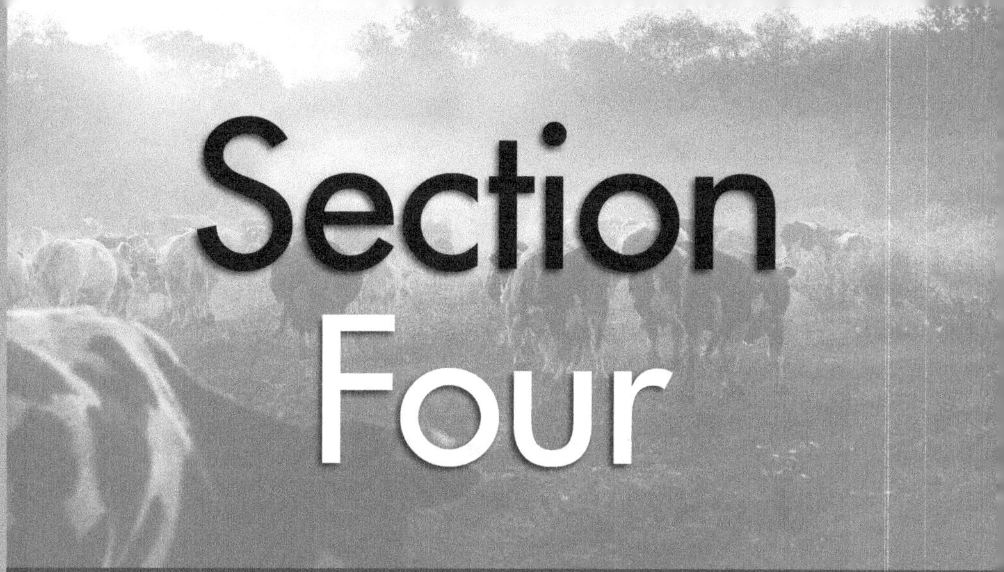

Section Four

Beef Cattle

Feel good studio/Shutterstock.com

Feel good studio/Shutterstock.com

Chapter 13

Breeds of Beef Cattle

Introduction

Many of the beef breeds used in modern agriculture had their origins in Europe. In the latter part of the 1700s, selection of cattle to form breeds began mainly in the British Isles. The cattle were selected for the most advantageous traits for the area. The most desirable animals were kept as breeding stock. The farmer then removed those animals that did not have the desired traits from the breeding herd. The practice of removing animals from the herd is referred to as culling. This method was repeated to increase gene frequency, and the early European farmers rarely used the technique except with breeding stock. As these purebred animals became popular, many breed registries were formed to establish the origins of the breed. Since then, all animals registered to the breed must trace the ancestry back to the original animals in the breed registry.

In the United States, some newer hybrid breeds have been developed as a result of selecting crossbred animals that possessed desirable traits. An example of this practice is the Brahman breed. Some other breeds that have been introduced in the United States include Simmental, Limousin, Chianina, Maine-Anjou, and Charolais, among others. More than 70 breeds of beef cattle are available to producers in the United States. Each breed has specific characteristics and origins that make it desirable for producers. Some points that should be considered when selecting a breed include the following: all breeds should have both strong and weak traits; no single breed is best for all traits; and every breed has a wide range of genetic variation. A farmer must keep in mind that when selecting animals as breeding stock, the use of good breeding practices has more impact than the particular breed selected. The producer needs to select a breed that seems to produce well in the area where it will be raised, and it should have proven market demand.

In this chapter, the focus will be on selecting beef cattle based on the specific goals of the cattle operation. Demonstrate a knowledge of the beef breeds and practice breed selection in Exercise 13-1.

Name____________________ Date__________

Exercise 13-1 Breed Selection Activity

When establishing a beef operation, a livestock producer must create goals for the farm and then decide which breed or crossbreed is best suited for the operation. For this activity, assume that the goals of the operation are (1) to have a low cost per unit of weight gain, (2) to have polled livestock to reduce herd injuries, (3) to produce a high-quality, well-marbled carcass, and (4) to have cows that will have a good mothering ability.

Directions: **Part 1:** Each of the following pictures show a different beef breed. For each picture, identify the breed and write a brief history and description of the breed. Include any unique characteristics of the breed. **Part 2:** Decide which breed or cross of breeds will meet the goals of the beef operation. Explain why the selected breed was the most advantageous.

Part 1

1.

© Red Angus Association of America

Figure 13-1

a. Breed: ____________________

b. History: ____________________

c. Description: ____________________

2.

Photo compliments of Beefmaster Breeders United, San Antonio, Texas

Figure 13-2

a. Breed: ____________________

b. History: ____________________

c. Description: ____________________

3.

Photo courtesy of Innisfail Farm

Figure 13-3

a. Breed: ____________________

b. History: ____________________

c. Description: ____________________

4.

American Simmental Association

Figure 13-4

a. Breed: ____________________

b. History: ____________________

c. Description: ____________________

Name______________________________ Date__________________

5. 

Lazarov/Shutterstock.com

Figure 13-5

a. Breed: ______________________

b. History: ______________________

c. Description: ______________________

6. 

American Brahman Breeders Association

Figure 13-6

a. Breed: ______________________

b. History: ______________________

c. Description: ______________________

Part 2

Decide which breeds or combination of breeds will meet the goals of the beef operation. Explain the decision in detail.

Name____________________________________ Date________________

Chapter 13 Matching Activity

Directions: Match the term to the corresponding definition or description.

Term	Definition
____ 1. gene pool	**a.** official record of registered animals
____ 2. trait	**b.** the range of different inherited traits
____ 3. cull	**c.** without horns
____ 4. Zebu breeds	**d.** distinguishing characteristics
____ 5. polled	**e.** the set of all genes, or genetic information, in a breeding population
____ 6. dual purpose	**f.** performance of the individual and its ancestors
____ 7. breed	**g.** preserved by cooling to very low temperatures
____ 8. herdbook	**h.** animals selectively bred for a long time and possess distinctive identity in size, conformation, function, and usually color
____ 9. dewlap	**i.** to remove from the breeding herd those animals that do not have the desired traits
____ 10. pedigree	**j.** physical appearance
____ 11. Wagyu beef	**k.** loose skin under the throat
____ 12. genetic diversity	**l.** renowned for its flavor and tenderness and is generally considered to be the highest quality beef in the world
____ 13. cryo-preservation	**m.** a breed of cattle from India that is a strain or hybrid related to the Brahman
____ 14. Conformation	**n.** used for milk and meat

Name____ Date____________________

Chapter 13 Lab Questions

1. When and where did beef cattle breed selection begin? Why?

2. What considerations are made when selecting breeds of cattle for a beef operation?

3. List the beef breeds that are most widely used in commercial beef production in the United States.

4. List beef cattle breeds that have been developed in the United States.

Name____ Date____________________

5. Why are genetic diversity and cryo-preservation important to the beef industry? Explain.

Feel good studio/Shutterstock.com

Chapter 14

Selection and Judging of Beef Cattle

Introduction

Beef cattle production is the most important agricultural enterprise in the United States. Cattle are adaptable to all regions of the country and produced in all 50 states. Income from beef cattle operations is of major economic importance at the national level as well as to individual producers. It is very important that producers understand the type of animals that are most efficient and most valuable in production. All traits of economic value should be considered when judging and selecting beef cattle.

Performance records help the producer select animals based on records of previous performance. Traits that are measured for performance are thought of as hereditable; therefore, a good performance score will increase the genetic value of the animal. Some traits are more heritable than others, so a cattle producer must use this knowledge to improve the beef herd.

Beef animals are judged on their conformation; performance records are given, and then they are placed on their visible body characteristics. Many county and state fairs will have cattle shows sometimes followed by a cattle sale, and FFA/4-H organizations have judging events, for exhibitors to promote their animals. Producers present their animals in a show ring to be judged in a class of similar cattle from other producers. The more judging events a beef animal wins or performs well in, the more valuable its genetics will be. A good judge learns the parts of the animal and develops a logical system for evaluating the animal. A judge should have experience and credibility in the cattle industry to build a reputation among cattle producers. Once a judge places a class of cattle in order from first to last, the judge should be prepared to explain why the cattle were selected in that order. The judge should give a presentation of oral reasons and use the proper terms to describe cattle conformation. There are cattle evaluation competitions in which FFA and 4-H members test their ability to properly select cattle based on conformation and breed standards.

This chapter provides the opportunity to practice livestock evaluation, which is useful for cattle producers, as well as an exercise to prepare oral reasons for a class of beef cattle.

Name________________________________ Date______________

Exercise 14-1 Beef Cattle Evaluation Scenario

Directions: Read the following scenario and use the expected progeny differences (EPDs) explained in the textbook chapter to answer the questions and place the class. EPDs are complex and you are encouraged to learn more about them; the below scenario only touches on the basics.

Note that in this scenario, the phrase "labor at calving is limited" means that there may not be someone available to help every calving heifer on the place. Therefore, high calving ease direct (CED) and low birth weight (BW) are important. You should also look for a high milk EPD so that the cows will be able to raise a healthy calf. Finally, the scenario indicates the operation will be selling bulls and marketing calves on the grid, so high scores in the remaining categories are also important.

Scenario

Twisted T Ranch is a purebred Gelbvieh operation that wants to start raising some balancer cattle. A few of the heifers will be sold to students to show and the rest will be kept as replacement heifers. The bulls will be sold at their annual sale and the rest of their calves will be marketed on the grid. This operation prioritizes sound cattle that excel in volume and fleshing ability. Feed and resources are adequate, but labor at calving is limited.

Number	Birthdate	CED	BW	WW	YW	Milk	REA	MARB
1	2/21/20	16	−1.0	49	98	20	0.70	0.50
2	2/18/20	9	1.7	85	131	16	0.60	0.40
3	3/1/20	14	0.0	74	122	22	0.89	0.62
4	2/25/20	6	2.1	60	90	12	0.35	0.30

CED: Calving Ease Direct
BW: Birth Weight
WW: Weaning Weight
YW: Yearling Weight
REA: Ribeye Area
MARB: Marbling

Questions

1. Based on the above scenario, which animal has the best data across the board? (Note that even though she does not necessarily have the highest calving ease or growth numbers, they are still high enough to work, especially since she has the carcass data, or REA and MARB, and milk to back it up.)

____

____

____

____

Name____ Date_____________________

2. Which animal has the best calving ease option but does not have impressive growth numbers?

____

____

____

____

3. Which heifer will likely raise the heavier weight calves but does not have the calving ease or birthweight to back it up? What would you do in this case?

____

____

____

____

4. The scenario says to prioritize calving ease because calving assistance is limited. With that being said, the scenario tells us to use heifer #___________ over heifer #___________. Explain why she would place above her.

____

____

____

____

5. Which heifer has the worst EPDs across the board?

____

____

____

____

6. Given the data, how would you place the four heifers?

____

____

____

____

Remember that EPDs are important, but calves must be as good in person as they are on paper. It is just as important to take their stoutness, soundness, and overall build into consideration as well.

Name______________________________________ Date____________________

Exercise 14-2 Giving Oral Reasons

Cattle evaluation is practiced in 4-H events, FFA competitions, and even at collegiate-level judging competitions. For this exercise, the instructor will provide access to a judging class of show cattle or you can look online. Use a group of approximately four animals for this judging exercise. While judging, make sure to view the animals from the front, side, and rear. If in person, walk around the cattle to see all the areas of interest for conformation and have the cattle move around the show ring as much as necessary for evaluation. Oral reasons will be given to explain the differences between animals that influenced their placings.

Use the following procedure to properly prepare to evaluate and give oral reasons for the group of cattle being evaluated.

1. Use proper terms that describe the characteristics of the animals. Beef cattle judging terms are listed in the textbook; use these terms when describing the differences in cattle.
2. When giving your reasons, stand 6 to 8 feet from the judge. Maintain eye contact and use good body posture during the presentation of reasons. Reasons are given in a logical order and should be well organized. Speak clearly and loudly enough to be heard. Speak with confidence and make a concise final statement explaining why the animal at the bottom of the class was so placed. Do not speak for more than 2 minutes total during reasons.
3. Keep accurate notes.
4. The evaluator should write notes in order to give accurate reasons. A set of reasons is organized in the following way:

   Introduction—Give the name of the class and how it was placed.

   Top pair—Give the reasons for placing 1 over 2, using comparative terms. Grant the advantages of 2 over 1, also using comparative terms. Finally, criticize 2 using either descriptive or comparative terms.

   Middle pair—Follow the same outline used for the top pair.

   Bottom pair—Follow the same outline used for the top pair.

   Summary—Give a final statement that explains why the animal at the bottom of the class was so placed. Mention the animal's strong points, if any, and describe its major faults.

Class #3: Market Steers		Placing: 3-1-2-4	
1 Black	2 Hereford	3 Black Baldy	4 Red
3/1	Heavy muscled More fat cover Apt to grade choice More lbs. of product Heavy boned	1/3 Grant Higher cutability Moderate frame	
		1 (Fault) Poor balance Over-finished	
1/2	Better off hind two legs Heavy muscled Market-ready look Wider chested	2/1 (Grant) Shapely, more expressive Clean lines Level hip	
		2 (Fault) Short coupled Narrow made	
2/4	Harvest with a more shapely carcass More even fat deposition	4/2 (Grant) Extended Structurally sound Deep bodied	
		4 (Fault) Light muscled Under finished Requires more days to grade choice Fine boned	

Figure 14-1

Name______________________________ Date______________

Chapter 14 Matching Activity

Directions: Match the term to the corresponding definition or description.

Term

____ 1. frame score
____ 2. growthy
____ 3. conformation
____ 4. ultrasonics
____ 5. performance testing
____ 6. production testing
____ 7. brood
____ 8. progeny testing
____ 9. 205-day weight
____ 10. pedigree
____ 11. sire summary
____ 12. maternal breeding value (MBV)
____ 13. expected progeny difference (EPD)
____ 14. type
____ 15. muscling
____ 16. finish
____ 17. carcass merit
____ 18. yield grade
____ 19. yield
____ 20. quality

Definition

a. the appearance of the live animal
b. the record of the ancestors of an animal
c. the refinement of the head, hide, bone, and hair
d. the general conformation of the animal; best determined from the side view
e. a measure of the degree of difference between the progeny of the bull and the progeny of the average bull of the breed in the trait being measured
f. an animal's own performance in relation to important economic traits
g. the amount of fat cover on the animal
h. the evaluation of a bull by the performance of its offspring
i. information on traits that are economically important to cattle producers
j. a tool to evaluate the mature skeletal size of an animal
k. the natural fleshing of the animal
l. a term used to describe an animal that is large and well developed for its age, is well balanced with adequate muscle, and has sufficient frame and bone structure to indicate good growth potential
m. the percent of carcass weight in boneless, closely trimmed retail cuts from the round, loin, rib, and chuck that have been closely trimmed of fat
n. a female animal that is kept for breeding in cattle
o. the kind of carcass the animal will produce when slaughtered
p. the use of high-frequency sound waves to measure fat thickness and loin-eye area
q. to measure a brood cow's production by the performance of its offspring
r. the dressing percent (weight of the chilled carcass compared to the live weight) of the animal
s. an estimate of the expected size of the animal when it reaches maturity; can be made from its frame score
t. ease of calving and ability to wean heavy calves

Name______________________________ Date______________

Chapter 14 Lab Questions

1. What are the main points to look for in judging market classes? Breeding classes?

2. What are the uses of beef performance testing records?

3. List the five points of organization to remember when giving a set of oral reasons.

Feel good studio/Shutterstock.com

Chapter 15

Feeding and Management of the Cow-Calf Herd

Introduction

Feeding programs for cow-calf beef herds are based on the use of roughages. Roughages are commonly provided by pasture in the summer and silage and hay in the winter. The kinds of pasture, silages, and hay used depend on the climate, soil, and land make-up of the geographic area. When preparing hay materials for cattle, dry hay can be produced in small or large square bales or, more commonly, in the form of round bales. Each category of beef animal has different feed requirements based on their stage of development. Dry pregnant cows are fed to prevent their becoming too fat or too thin during the winter. Younger cows and heifers require more feed than mature cows. Breeding bulls also have periods of high energy requirements, such as during the breeding season. Calves may be creep fed commercial pelleted grain when being marketed after weaning.

Bull calves to be kept for breeding purposes are weaned at 45 to 90 days. High-energy rations are fed for five months to determine which bulls have the best gaining ability. Bulls that gain best should be selected for breeding, and the others should be marketed for slaughter. The bulls that are kept should begin breeding at 15 to 18 months of age. With the proper breeding strategy, calving should occur in a 40- to 60-day period. Keeping births within this time period gives calves a more uniform weight and age. Calving should be timed out to occur six to twelve weeks before pasture season, depending on local climate.

During calving, producers should monitor the cows closely for signs of parturition. Most cattle on pasture will not need assistance in delivery, but many young heifers may need help during calving. After birth, calves should be maintained with common herd health practices. Castration, dehorning, vaccination, and identification of calves are best done when they are young so there is less shock to the calf since its sensory neurons are not yet fully developed. Some farmers who do not own cow herds may buy calves at weaning and prepare them for the feedlot, where they will be finished and sent to the slaughterhouse.

This chapter analyzes the feeding requirements of beef cattle at different stages of production in Exercise 15-1. Exercise 15-2 helps to develop proper calf care practices.

Figure 15-1 Feed bunk on a large feedlot.

Name____ Date____________________

Exercise 15-1 Feeding Cows and Calves at Different Stages of Growth

Maintaining feed with the proper available energy for the requirements of beef cattle at different stages of growth is imperative efficient management. Producers try to avoid cattle losing too much condition or gaining too much condition based on the stage of development of each animal.

Directions: For each of the following categories of beef cattle, describe the suggested ration for proper body condition maintenance. (*Hint:* Suggested rations are provided in the textbook.)

1. **Creep Feeding Calves—Self-Fed (100-lb mix)**

____

____

____

____

____

2. **Feed Mixtures for Weaned Bull Calves (weaning to about 700 lb)**

____

____

____

____

____

3. **Feed Mixtures for Bulls over 700 lb**

____

____

____

____

____

4. **Wintering Rations for Bred Heifers (800 to 900 lb)**

____

____

____

____

____

Name____________________________________ Date________________

5. **Rations to Grow Replacement Heifers (450–500 lb; weight gain 1–1.25 lb daily)**

6. **First-Calf Heifers—Lactation Rations—Drylot**

7. **Lactating Rations for Cows in Drylot (1,100 lb)**

8. **Wintering Dry, Pregnant Beef Cows (1,000 to 1,100 lb)**

Name______________________________ Date______________

Exercise 15-2 Calf Care Activity

Calf care begins from the time that a cow begins the calving process. A cattle manager must understand the signs and steps in the calving process to assist in a safe and healthy calving experience. Once the calf is born, several immediate practices are necessary to help the calf to survive the first 48 hours of life. Once a calf has passed that threshold, other management practices are needed to prepare the calf for the purpose of the beef operation.

Directions: For each of the calf care practices mentioned, describe the proper implementation of the practice in detail.

Calving practices and procedures:

____

____

____

____

____

First 48 hours after calving:

____

____

____

____

____

Castration:

____

____

____

____

____

Dehorning:

____

____

____

____

____

Name____ Date__________________

Branding and marking:

Weaning:

Managing calves after weaning:

Name______________________________ Date______________

Chapter 15 Matching Activity

Directions: Match the term to the corresponding definition or description.

Term

____ 1. puberty
____ 2. feed composition table
____ 3. rotation grazing
____ 4. carrying capacity
____ 5. husklage
____ 6. creep feeding
____ 7. artificial insemination
____ 8. castration
____ 9. preconditioning
____ 10. backgrounding
____ 11. synchronization of estrus
____ 12. scours
____ 13. placentophagia
____ 14. steer
____ 15. staggy
____ 16. burdizzo
____ 17. elastrator bands

Definition

a. the number of animals that can be grazed on the pasture during the grazing season
b. gradually increasing the amount of grain provided for a calf as it grows
c. the growing and feeding of calves from weaning until they are ready to enter the feedlot
d. the removal of the testicles of the bull calves
e. a management tool available to beef producers that uses a team approach to improve the competitiveness, efficiency, and profitability of their beef business
f. the process of preparing calves for the stress of being moved into the feedlot
g. the analysis of feeds that can be used for rations for beef herds
h. the material that comes from the corn combine that can be used as a feed
i. the division of a field using temporary fencing when feeding cattle
j. the placing of sperm in the female reproductive tract by other than natural means
k. the age at which heifers come into heat
l. special instrument used to place a tight rubber band around the scrotum above the testicles
m. bulls that are castrated after eight months of age
n. castrated male beef animal
o. the use of various compounds (usually hormones) to cause all females in a herd to come into heat within a short period of time
p. act of the mother eating her placenta, also known as afterbirth
q. diarrhea

Name______________________________ Date________________

Chapter 15 Lab Questions

1. Why is it necessary to adjust the rations fed to bulls and cows during and after breeding season?

2. What are the advantages and disadvantages of using artificial insemination with a beef breeding herd?

3. Explain precondition and backgrounding when raising calves and why it is important.

Feel good studio/Shutterstock.com

Feeding and Management of Feeder Cattle

Introduction

Most cattle feeding operations exist in the grain-producing states of the north-central and southern regions of the United States. Beef cattle have high grain requirements for finishing; therefore, many cattle operations are established in or near the grain-producing states. The large commercial feedlots raising more than 1,000 head of cattle are found mostly in the Plains states. The smaller feedlots of less than 1,000 head are found mainly in the Corn Belt and other locations across the United States.

Feeders are cattle that will be raised based on their daily rate of weight gain. Producers of feeder cattle have a goal of feeding each animal enough to make them grow at an appropriate daily rate so that the cost of raising the animals is less than the return gained by selling them. Some feeders use more roughage in the ration under a system of deferred finishing. Yearling feeders may be fed crop residues before being put on a full feed to clean up otherwise unused feed. Steers have higher weight gains in a feedlot than heifers. Calves make more efficient daily weight gains than older cattle, but they take longer to reach market weight. Crossbred feeder cattle have proven to gain weight faster than the pure parent breeds. More roughages can be used when feeding lower grades of feeder cattle because they are not producing high rates of gain.

Feeder cattle are purchased by the producer rather than using a cow-calf operation. Most feeders come from cattle auctions or are sold directly from the farm or ranch where they were born. The highest level of feeder cattle sales takes place during the fall months. Many producers sell their feeders at this time because, at this point, the productive season of pasture is winding down, the weather is getting colder, and the rate of gain will decrease while the cost of feed will increase. Using accurate records, a farmer can calculate the profit from feeding cattle. The profit is equal to the value of the finished cattle minus the total cost.

In the United States, most feeder cattle are fed using corn; corn silage is the best roughage that can be used for increased rate of weight gain. For the protein requirements in a feeder cattle ration, many natural proteins are used, such as cottonseed meal, soybeans, and many others.

Other dietary requirements for feeders are salt, calcium, and phosphorus. It is sometimes necessary to add vitamins A, D, and E when finishing cattle.

One measure of feed efficiency is the feed intake of the cattle. Feed intake is influenced by the energy level in the ration, weather, feed palatability, feed processing, and the degree of finish on cattle. When starting cattle on feed, the proportion of roughage in the ration should be higher at the start and then gradually be decreased as the animal grows and moves onto a full grain feed. The proportion of concentrate is gradually increased until the cattle are on full feed. The amount of roughages used before going onto full feed ranges from 10 to 70 percent; the proportion of feed not provided by roughages will contain concentrates.

Because new cattle come on and off the farm often in a feeder cattle operation, the cattle must be watched closely for signs of sickness and disease. New cattle should be isolated in small lots with fresh feed and water. Vaccination and parasite control programs must be carefully planned and implemented soon after new cattle arrive. In this chapter, complete the feeder calculation exercise. It is good practice to plan the feeding regimen of various types and sizes of feeder cattle.

$$\textbf{Break-even feeder price} = \frac{\textbf{fed weight}}{\textbf{feeder weight}} \times \textbf{expected price of fed cattle} - \frac{\textbf{gained weight}}{\textbf{feeder weight}} \times \textbf{cost of gain}$$

Name____ Date____________________

Exercise 16-1 Feeder Cattle Calculation Lab

Beef cattle producers depend on every dollar per hundredweight of price for their animals. Usually, producers will use a break-even analysis to measure their returns using the current market price and the current cost of production. Several cost-versus-return calculators are available online for beef producers to accurately estimate their profits or losses.

Directions: Refer to the following sample values for feeder cattle costs and returns and determine the following performance measures using the values given in Table 16-1.

Cost of Interest

Average daily gain (deads out)
Average daily gain (deads in)
Total pounds of feed, as fed
Feed efficiency (lb of feed/1 lb of gain—deads accounted for)
Cost of gain (without return to management—deads accounted for)

1. Return to ownership and management

____

____

____

2. Appropriate purchase price

____

____

____

3. Sale price to attain given return

____

____

____

Name________________________________ Date____________

Table 16-1 **Sample Feeder Cattle Costs**

Price paid per pound ($)	$0.8600
Purchase weight (lb)	600.00
Trucking fee—arrival ($)	$3.50
Hired labor cost per head ($)	$2.00
Annual interest rate (that is 10)	10.00%
Price paid per pound of diet, as fed ($)	$0.050
Pounds of diet fed daily, as fed	10.00
Days owned	222.00
Yardage per day ($)	$0.29
Cost for processing products ($ for vaccines, implants, dewormers, tags)	$6.50
Morbidity rate (%)	5.00%
Cost of treatment ($) (entire cost of multiple-day treatment)	$12.00
Re-pull rate (%)	11.00%
Cost of re-pull O34 treatment ($)	$16.00
Mortality rate (%)	1.50%
Chronic rate	1.00%
Price received at sale ($/lb)	$0.9800
Unshrunk sale weight (lb)	1264.53
Percent shrink (%)	3.00%
Sales or commission fees ($)	$1.00
Trucking fee—departure ($)	$4.00
Decreased return for chronics (compared to pen average)	$130.00

Name____ Date____________________

Chapter 16 Matching Activity

Directions: Match the term to the corresponding definition or description.

Terms

____ 1. efficiency in gain
____ 2. haylage
____ 3. boot stage
____ 4. high-moisture storage
____ 5. rolling
____ 6. crimping
____ 7. pelleting
____ 8. middling
____ 9. NEg
____ 10. bacterin
____ 11. grass fed
____ 12. finishing
____ 13. concentrates
____ 14. stillage
____ 15. grinding
____ 16. cubing
____ 17. lignified

Definition

a. refers to harvesting the grain at a high moisture content (22 to 30 percent for corn and milo) and storing it in a silo
b. the same as rolling, except the rollers used have corrugated surfaces
c. refers to grinding the feed into small particles and then forming them into a small, hard form called a pellet
d. net energy for gain on a dry-matter basis
e. refers to the amount of feed needed for each pound (kilogram) of gain
f. a liquid containing dead or weakened bacteria that causes the animal to build up antibodies that fight the disease caused by the bacteria
g. that point in growth at which the inflorescence (the flowering part of the plant) expands the boot
h. refers to processing the grain through a set of smooth rollers that are set close together; the grain is pressed into the form of a flake
i. low-moisture grass silage
j. a by-product of the flour milling industry; provides protein and energy in the ration and contains a good supply of phosphorus and potassium
k. grass and forage shall be the feed source consumed for the lifetime of the ruminant animal; animals cannot be fed grain or grain by-products and must have continuous access to pasture during the growing season
l. woody by the deposition of lignin in cell walls
m. the same process as pelleting, but are larger in size than the pellets
n. the process of bringing cattle to full weight prior to harvesting them for beef
o. distiller's grains are a by-product of grain alcohol fermentation
p. processing a feed through a hammer mill
q. low in fiber but rich in energy and protein

Name_______________________________________ Date_________________

Chapter 16 Lab Questions

1. Where are most feeder cattle operation located? Why?

2. At what time of year are the most feeder cattle sold? Why?

3. What factors may affect feed intake?

Feel good studio/Shutterstock.com

Chapter 17

Diseases and Parasites of Beef Cattle

Introduction

When cattle are raised in confinement such as a beef lot, the presence of various diseases, bacteria, and parasites makes herd health management an integral part of the operation. Implementing a good herd health management program is essential because when cattle get sick, the profitability of those animals decreases. With the assistance of a veterinarian, a beef producer should create a herd health prevention plan that will include several strategies to reduce the risk of a disease or parasite spreading. These strategies may focus on changing environmental factors, conducting routine herd testing, separating groups of cattle, and even culling infected animals.

Vaccination is a tool routinely used by beef producers to reduce the threat of disease. Antibiotics and sulfa drugs are used to treat a multitude of diseases and infections. The concern is that large-scale use of certain antibiotics may lead to strains of the disease that are resistant to the medicine. When beef producers buy replacement cattle, they should acquire cattle from disease-free herds and always isolate new animals for a period of time to help in disease control. Many diseases display similar symptoms, which can result in misdiagnosis. Laboratory tests and the aid of a veterinarian are often needed to accurately identify the disease that is present.

Insects spread many diseases, and their control is just as important in preventing the spread of disease in cattle. The most common external parasites of cattle are flies, lice, mites, and ticks. Some cattle operations use insecticide to control most of these parasites. All insecticide material must be labeled with directions that need to be followed carefully for safe use. Roundworms, flatworms, coccidia, and anaplasma are the most common internal parasites of beef cattle. When treating for internal parasites, sanitation is the most effective control.

Beef cattle are at risk of contracting many health problems, and several modes of infection are possible. Vaccination, sanitary environment, and proper management of feeding programs help prevent some of these problems. Exercise 17-1 provides practice in identifying symptoms and implementing proper prevention and treatment programs.

Name______________________________ Date______________

Exercise 17-1 Disease Prevention and Treatment Activity

Directions: For each of the following diseases or sicknesses that can affect beef cattle, identify the source of the disease, describe the symptoms that are displayed, and recommend common prevention and treatment methods.

Disease/Sickness	Sources of Disease	Symptoms	Prevention/Treatment
Bovine spongiform encephalopathy (BSE)			
Bovine virus diarrhea (BVD)			
Brucellosis			
Blackleg			
Foot and mouth disease			
Johne's disease (paratuberculosis)			
Leptospirosis			
Pinkeye (infectious keratitis)			
Shipping fever (*Pasteurella*)			
Campylobacteriosis			

Name____ Date____________________

Chapter 17 Lab Questions

1. Which disease(s) may not show many symptoms until some of the cattle die suddenly?

2. What factors should a herd health plan address?

3. When buying replacement cattle, what steps should a producer take to reduce the spread of disease?

4. List three diseases that have one similar symptom.

5. Describe three ways to reduce the amount of insects around cattle.

Feel good studio/Shutterstock.com

Chapter 18

Beef Cattle Housing and Equipment

Introduction

When building a new cattle operation or expanding an established farm, detailed planning is necessary to avoid costly mistakes or setbacks. Well-planned facilities save labor and make the operation more efficient in many ways. When designing a facility, the producer should address issues such as employee and animal safety, allow for efficient animal handling, and reduce sources of stress for the animal. When planning a new facility, always plan for possible expansion in the future as the business grows. In general, cow-calf herds have the simplest facility requirements. The cows can calve on pasture, and in the winter months, some farmers will provide open-faced barns for shelter. Feedlot operations can range from widely complex to extremely simple, depending on the size and scope. Many feedlots use some form of automation during production to decrease labor costs.

When building confinement housing, warm barns do not make sense for a beef operation because they do not increase the rate of gain. For beef cattle, cold confinement barns provide necessary shelter from the elements. Many of these confinement buildings are being built with slotted floors to make manure handling more efficient and to reduce the amount of manure in the barn. When designing a feedlot, windbreaks should be constructed to protect cattle from the weather.

Properly designed corral systems save beef producers a lot of time, reduce stress on the animal, and reduce injuries to cattle and humans. Researchers have recently made advances in understanding the factors that increase stress in corral systems. Corrals provide a place to sort and work with the cattle for herd checks and when treatments are needed; they also make it easier and safer to handle cattle. A corral should include holding pens, sorting pens, alleyways, a working chute, a squeeze chute, a headgate, and a loading chute. A cattle farmer may want to use a scale for accurate measurements of rate of gain and to determine the "on the hoof" weights of cattle being sold.

Feeding and watering facilities should be planned carefully as well. Feed storage facilities need to be designed; feed is stored in silos and feed bins. Feed bunks of various kinds are used to provide a controlled amount of feed to the cattle. Portable bunks are the least expensive and most commonly used. High-tech mechanical feed bunks save labor but require a greater initial investment in equipment. An adequate supply of clean water is extremely important for a beef cattle facility. Automatic waterers are commonly used to reduce labor costs. Waterers need to be routinely cleaned to maintain sanitation. Other equipment used in beef operations includes mineral feeders, self-feeders, hay rings, creep feeders, and many other types of equipment to improve efficiency.

Name____ Date____________________

Exercise 18-1 Design a Modern Cattle-Handling Facility Lab

When designing cattle-handling facilities, careful attention must be given to make the facilities effective to move, sort, and handle cattle safely. A curved corral design is usually beneficial when gathering and sorting beef cattle. There are several common designs for cattle corrals to avoid injury to the animals and to the cattle handlers. Figures 18-1 through 18-3 give examples of common facilities for reference.

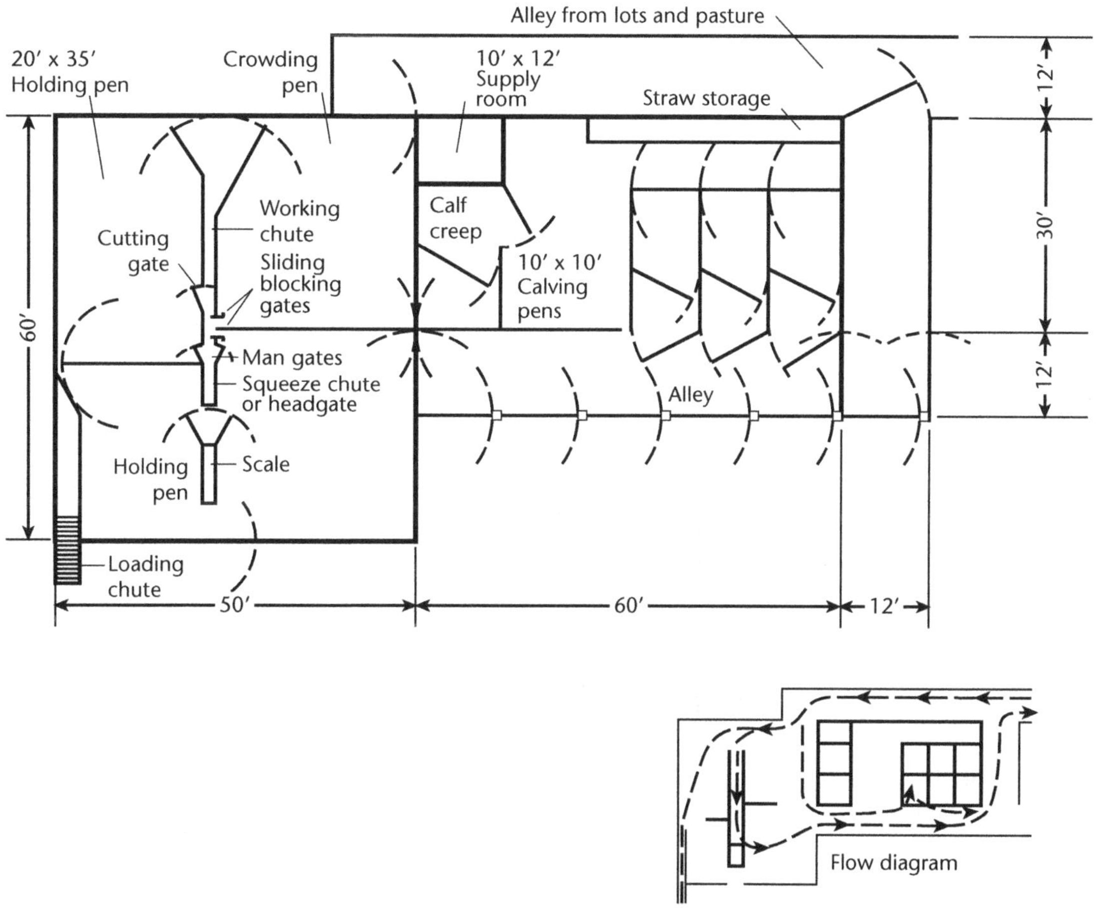

Figure 18-1 Plan for a corral for 300 to 1,000 head of cattle.

Name____ Date____________________

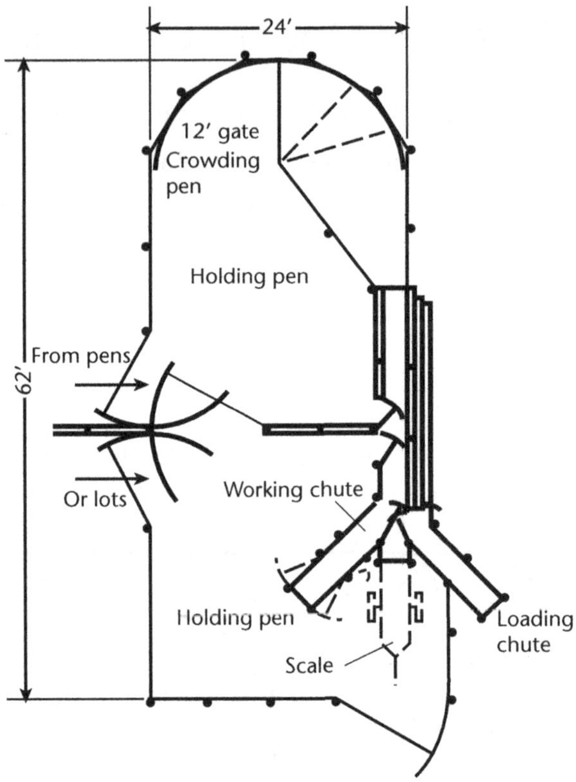

Figure 18-2 Two plans for small herds up to about 75 head.

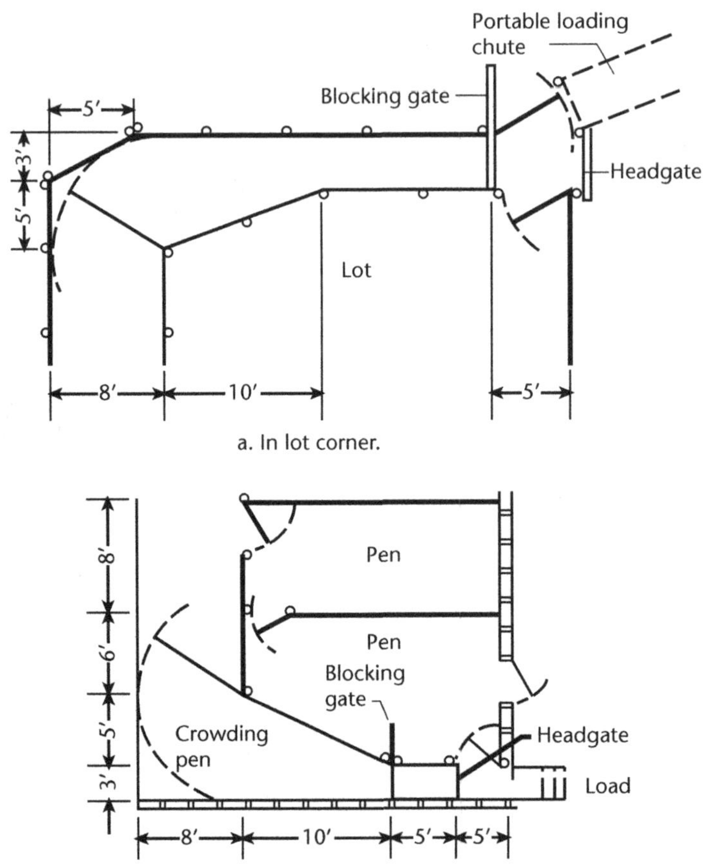

Figure 18-3 Half-circle chute. The entrance into a curved corral.

Name____ Date____________________

Directions: Create a design drawing for a cattle corral and handling chute. The drawing should be detailed and contain straight, clean lines. The drawing should be thoroughly labeled and include all dimensions. Use the working facility dimensions in Table 18-1 to make sure the design is appropriate for 600- to 1,200-lb beef cattle.

Table 18-1

Corral and Working Facility Dimensions

	To 600 lb	**600–1,200**	**<1,200 and Cow-Calf**
Pen Space (sq ft/head)	14	17	20
Crowding Tub (sq ft/head)	6	10	12
Working Chute-vertical sides			
Width (inches)	18	20–24	26–30
Minimum Length (feet)	20	20	20
Working Chute-sloping sides			
Width at Bottom (inches)	13	15	16
Width at Top (inches)	20	24	28
Minimum Length (feet)	20	20	20
Working Chute Fence			
Height—minimum	45	50	60
Depth of Posts—minimum	30	30	30
Corral Fence			
Height	60	60	60
Depth of Posts—minimum	30	30	30
Loading Chute			
Width (inches)	26	26	26–30
Length (minimum, feet)	12	12	12
Rise, in/ft	3½	3½	3½

Dimensions from *Corral and Working Facilities for Beef Cattle.* GPE-5002

Create the design drawing in the following space (use a separate piece of paper if needed):

Name____ Date____________________

Exercise 18-2 Equipment Cost Lab

When designing a cattle operation, one of the major limiting factors is cost. Some operations begin with a large amount of capital and are able to purchase all types of handling equipment, but many young start-up farmers do not have that kind of money available. This exercise is a class competition. The goal of the competition is to create the most cost-efficient facilities for a cattle operation. The design must be low cost, but it must also provide all necessary equipment for cattle to grow comfortably and facilities to handle the cattle. The student(s) who designs the most cost-efficient cattle farm will win the challenge.

Directions: A list of available equipment is provided here. Design and sketch a complete beef cattle facility. Include all the necessary facilities and equipment to feed and grow cattle, as well as to load and unload cattle for transport. Create a cost analysis of the facility and turn it in to the instructor along with the sketch of the facilities. The student who has the lowest cost analysis while still maintaining a functional operation will win the competition.

Equipment Price List

Heated confinement barn for 500 head—$400,000
Cold confinement barn for 500 head—$250,000
Pour slab and cement curbs for outdoor feedlot for 500 head—$120,000
Feedlot fencing for 500 head—$105,000
Foundation fill (topsoil and stone) brought in as an alternative to cement—$8,000
Slotted floors—$20,000
Sub-floor manure lagoon—$80,000
Outdoor manure lagoon—$50,000
Bunk silo—$75,000
Upright silo—$100,000
Grain bins for 500 head—$25,000
Portable mixer/grinder—50,000
Sunshade for open feedlot for 500 head—$20,000
Open-face run-in shed (each holds 100 head)—$10,000
Back rubbers (each one services 50 head of cattle)—$2,000
Creep feeders (25 animals each)—$1,000
Self-feeders (25 animals each)—$5,000
Cement bunk feeders for 500 head—$15,000
Calf shelters (50 animals each)—$5,000
Cattle-guard gates—$1,500
Cattle treatment stall—$4,000

Name____ Date____________________

Windbreak for feedlot—$20,000

Corral system for 500 head—$15,000

Headlock gate—$2,000

Headlock chute—$5,000

Headlock chute with tilting table—$10,000

Automatic manure scrapers for allies for 500 head—$30,000

Surveillance camera system for management—$5,000

Directions: Give this page to the instructor for competition evaluation.

Sketch of cattle facilities in the following space:

Cost analysis of facilities:

Name____ Date____________________

Chapter 18 Matching Activity

Directions: Match the term to the corresponding definition or description.

Term

____ 1. headgate
____ 2. tilting table
____ 3. dipping vat
____ 4. cattle guard
____ 5. stocking rate
____ 6. carrying capacity

Definition

a. a device for restraining the animal in a horizontal position while it is being treated
b. allows equipment to be driven into an area without the operator having to stop and open a gate
c. used to hold the cattle while certain treatments are performed
d. used to treat livestock for pest control
e. the number of animals that can be grazed on the land for one season
f. the concentration of grazing livestock on a given amount of land over a period of time

Name______________________________ Date______________

Chapter 18 Lab Questions

1. What factors should be considered when planning a cattle facility?

2. What type of operation has the lowest cost? Why?

3. Why is new technology not always the best option when building cattle facilities?

feel good studio/Shutterstock.com

Chapter 19

Beef Cattle Marketing and Management

Introduction

There are several options for conventional beef cattle producers to market or sell their product. There are also some alternative markets for beef. The price that is paid for beef cattle is controlled by supply and demand. Producers have no control over demand—that factor is consumer-driven. Producers do have some control over the supply, but other factors such as extreme weather conditions and available feed affect the supply of cattle as well. Beef cattle prices vary with the season of the year. Long-term trends or cycles in beef cattle prices have proven to drive price also.

There are three common methods by which beef cattle are marketed: auction markets, terminal markets, and direct selling.

1. Auction Terminal markets are also called public stockyards or sale barns found in many local communities; the facilities of the auction terminal market are owned by a stockyard company, and the cattle remain the property of the sellers for the and buyers. The stockyard company charges a fee for selling the animal, the use of the facilities, and for feed fed to the cattle while they are in the stockyard. The cattle are marketed through the auction and are sold to the highest bidder in person or even online based on the weight of the animal and market prices. The importance of terminal markets has been declining in recent years because there are several other more convenient options for producers.
2. Terminal markets are cattle sold by commission agents to buyers who, in the agents judgment, will be the strongest bidders. Auction markets are found in many local communities and are also called sale barns. The numbers of cattle marketed in auction markets have not changed much over the years.
3. More cattle are being sold by direct marketing methods now than they were a few years ago. There are several forms of direct marketing. Contract sales are used to market cattle directly from feeder cattle producers to cattle feeders. Cattle can be bought and sold by a cattle dealer or to order buyers that buy feeder cattle on order for cattle feeder operations. Online and

electronic marketing have advanced in recent years, and a buyers can now view and bid on cattle on their home computer systems or cell phones.

Purebred producers and value-added producers commonly use other direct marketing options such as private sales. In these marketing situations, the reputation of the breeder and the methods used in production and advertising are important factors in the sale of these specialty cattle. The sale of beef cattle that are transported across state lines is regulated by the Packers and Stockyards Act. This act created rules for fair business practices and competition in the marketplace, and it is enforced by the U.S. Department of Agriculture (USDA). Cattle shrink in weight when being shipped to market. Careful handling and management can reduce shrinkage loss and carcass damage that may occur during shipping.

Many producers select a marketing method based on convenience, but price and costs of marketing are also important factors that need to be considered. During sales, cattle are divided into classes based on use, age, weight, and sex.

Figure 19-1 Beef carcass during processing.

Feeder and slaughter cattle are given a grade to describe their body condition and conformation. Quality grade is determined by the age of the animal and the amount of marbling in the carcass. The yield grade is determined by the amount of lean meat that can be cut from the carcass. The grade given will affect the price paid for the animal.

Another marketing option for producers is to use the cattle futures market to reduce price risk. A price is contracted based on futures market predictions, and the buyer and seller are locked in at that price for the duration of the contract. If the market prices drop, the producer's price will remain the same, but the producer will not see added profits if the market price increases. A cattle producer must keep good records and have an understanding of the pricing system to successfully use the cattle futures market.

In this chapter, the exercises will explore beef marketing methods and alternative markets.

Name________________________________ Date________________

Exercise 19-1 Marketing Methods Lab

A producer must choose a market based on the best options for their operation. A choice of market depends on many factors, such as price, size of animal, convenience for producer, and reasons for sale.

Directions: For each of the following market descriptions, determine if it is a terminal market, an auction market, or direct selling.

1. The Angus association is hosting an annual feeder cattle sale. The marketing costs include charges for yardage, feed, insurance, brand, and health inspection. Charges are based either on a percentage of the selling price or a fixed fee.

   Marketing method: ________________________________

2. The sale is made on the range or farm where the feeder cattle are produced. The producer enters a contract with the party that is buying the feeder cattle.

   Marketing method: ________________________________

3. The producer consigns cattle to a commission firm. The firm does not take title of the cattle, but takes a fee to sort animals into uniform lots for sale and carry out the sale.

   Marketing method: ________________________________

4. A cattle buyer uses an Internet auction program to see cattle that will be for sale in an online auction. The buyer views the cattle from video footage available online, and then enters a bid on the cattle.

   Marketing method: ________________________________

5. About 12 percent of the meat packer sources of slaughter cattle are purchased from this type of cattle market.

   Marketing method: ________________________________

6. An order buyer is a broker that buys slaughter cattle for packing plants. The cattle are bought on the farm and shipped directly to the packing plant.

   Marketing method: ________________________________

Name__________________________ Date__________

Exercise 19-2 Alternative Beef Markets Activity

Some beef marketing strategies have unique factors and demand a higher price than conventional market options. The following is an example of a small-scale beef producer with value-added factors in their beef herd.

Sample Beef Cattle Operation

This beef operation consists of a cow-calf operation that raises the male calves for feeder stock and heifers as replacement heifers to grow the herd. The herd is purebred Angus cattle from top-grade genetics. The herd is also grassfed to add market value. The producer wants to get the top dollar value for the cattle but also considers convenience an important factor.

Directions: As a hired marketing consultant, choose an alternative market for the sample beef cattle operation. Describe how the producer will make a deal and transport the cattle as well as how the product will get to the end customers. Give the details as to why this is the best option for this farm. Compare it with other alternatives.

Alternative Market Analysis: (describe the suggested marketing plan here)

Name______________________________ Date______________

Chapter 19 Matching Activity

Directions: Match the term to the corresponding definition or description.

Terms

____ 1. supply
____ 2. demand
____ 3. stockyards
____ 4. commission
____ 5. yardage
____ 6. auction markets
____ 7. fill back period
____ 8. slaughter calves
____ 9. bullock
____ 10. quality grade
____ 11. vealers
____ 12. retained ownership
____ 13. expected progeny difference
____ 14. feeder calves
____ 15. yearling feeders
____ 16. dark cutting beef
____ 17. Packers and Stockyards Act
____ 18. checkoff
____ 19. hot carcass weight
____ 20. marbling

Definition

a. a fee for a firm's services in marketing the cattle

b. usually cattle between three and eight months of age that have had feed other than milk in their diet for a period of time

c. the amount of a product that buyers will purchase at a given time for a given price

d. a male, usually under 24 months of age, that may be castrated or uncastrated and does show some of the characteristics of a bull

e. option where the facilities of the terminal market are owned by a stockyard company

f. defined as the amount of a product that producers will offer for sale at a given price at a given time

g. based on the amount and distribution of finish on the animal

h. the fee charged for the use of stockyard facilities

i. time during which cattle are fed after the cattle reach market

j. where cattle are sold by public bidding, with the animals going to the highest bidder

k. is the unchilled weight of the carcass after harvest and the removal of the head, hide, intestinal tract, and internal organs

l. feedlot arrangement in which cow-calf or stocker operators retain ownership of the cattle until they are harvested

m. calves that are one to two years of age and are sold to be fed to finish for slaughter

n. a $1 per head money that is collected and used to promote beef as a food, provide consumer information about beef, provide information to the beef industry, promote foreign marketing, and promote research in beef production and utilization

o. calves grown for veal

p. the presence and distribution of fat and lean in a cut of meat

q. weaned calves that are under one year of age

r. an estimate of the genetic value of an animal in passing genetic traits to its offspring

s. a federal law that is administered by the USDA; all cattle that move across state lines are regulated by this act

t. a condition in which the lean meat is darker than normal in color

Name____ Date____________________

Chapter 19 Lab Questions

1. What is the beef checkoff program? Describe.

2. How does the pork supply affect cattle prices?

3. Describe the different grades a beef carcass can receive. How does this affect price?

4. What is the Packers and Stockyards Act? How is it enforced?

Section Five

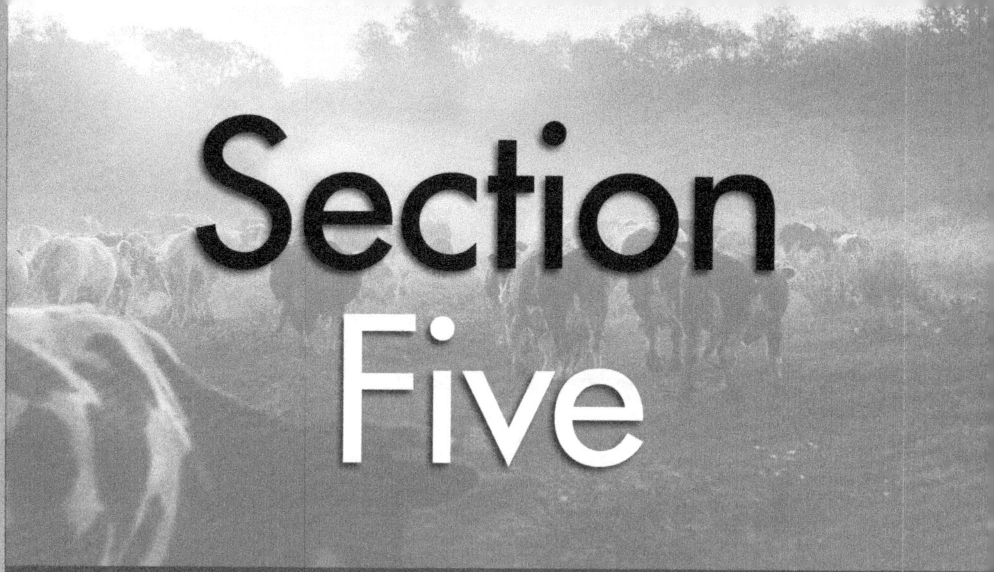

Swine

Feel good studio/Shutterstock.com

Feel good studio/Shutterstock.com

Chapter 20

Breeds of Swine

Introduction

The major part of the swine ration is grain; this is why many swine operations in the United States are located in states where feed grain is produced in large quantities. More than half of the hogs raised in the United States are raised in the midwestern states. Swine are efficient at converting feed into meat; therefore, they have been a popular choice of livestock. Some swine producers jump in the market and buy pigs when the sale price is high and then leave the business when the prices are in a low trend. This practice has become less common recently because many producers have invested in modern technology, equipment, and facilities, thus making it more economical to continuously raise swine.

Hog producers often tend to use crossbreeds because they find many advantages to raising good crossbred swine. Selection of breeds to use in crossbreeding programs is based on data measuring litter size, growth rate, feed efficiency, carcass length, leanness, and muscle. The market demand for carcass quality pushes producers to manage many factors, from breeding to environment.

Of the many swine breeds in existence in North America, several also have their origins in the United States. Other breeds were developed in Denmark and England. Just as in other livestock industries, purebred swine breeds have associations that register only purebred animals. Standards for registration of the individual breeds are established by the breed associations. Purebred swine are important for crossbreeding as well. Without the purebred breeding stock, the crossbred qualities would be diluted.

Controlled inbreeding has been developed and tested in the United States by state experiment stations as well as by private breeders. Inbred lines of hogs have been used in crossbreeding programs to improve litter size and growth rate. Another reproductive strategy becoming more relevant in swine production is genetic engineering. Genetic engineering holds promise of improving swine breed characteristics beyond what traditional breeding strategies have accomplished. This chapter develops a solid understanding of the common breeds of swine in the United States.

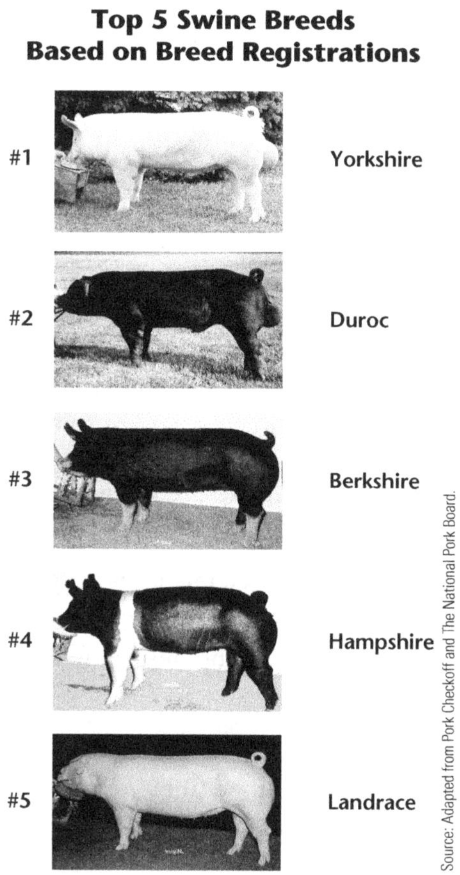

Figure 20-1 The top five most popular swine breeds based on breed registrations.

Name____ Date____________________

Exercise 20-1 Breed History and Function Activity

In the modern swine industry, the performance and growth of pigs are measured more than ever. Only producers who intensively manage their herds are successful. Understanding the characteristics of different breeds is important knowledge for a pig farmer.

Directions: For each swine breed description, identify the matching pig breed. Answer all of the breed questions at the end of the exercise using the textbook.

1. Breed:____________________

   These black-bodied pigs have six white points, including their nose, tail, and feet. They have erect shorts ears and dished snouts. Legend says this breed was discovered by Oliver Cromwell's Army at Reading (the country seat of the shire of Berks) in England more than 300 years ago. They are known for providing hams and bacon of excellent flavor. They were first brought to America in 1823. Much improvement has been made through testing and genetic evaluation to meet the demand for fast, efficient growth, reproductive efficiency, and leanness.

2. Breed: ____________________

   Known as the durable mother breed, these pigs have white bodies with long, droopy ears. This breed originated in Chester County, Pennsylvania. More than 60,000 animals are recorded by this breed's association each year. They have maintained their popularity with pork producers because of their mothering ability, durability, and soundness.

3. Breed: ____________________

   These red pigs with drooping ears are the second most recorded breed of swine in the United States and a major breed in many other countries. Their color can range from very light golden brown—almost yellow—to a very dark red that approaches mahogany. The growth of the breed is, in part, due to characteristics such as the ability to produce large litters, longevity in the female line, lean gain efficiently, carcass yield, and product quality as a terminal sire.

4. Breed: ____________________

   These black hogs have white belts across the shoulders, covering the front legs around the body. They have erect ears and are heavily muscled. They are the third most recorded swine breed in the United States. They are leaders in leanness and muscle, with good carcass quality, minimal amounts of back fat, and large loin eyes. Their ability to produce winning carcasses is unequaled, and they continue to set the standard by which all other terminal sires are evaluated. The females are known as great mothers and excellent pig raisers and have extra longevity in the sow herd.

5. Breed: ____________________

   These white pigs with large drooping ears are the fourth most recorded breed in the United States as well as a major breed in many other countries. Their purebred females are known for their ability to produce large litters over an extended time. Boars are aggressive and sire large litters that combine growth, leanness, and other desirable carcass traits. This, along with their outstanding maternal traits, has made them leaders in swine operations throughout the world.

6. Breed: ____________________

   These black-and-white–bodied pigs may have a white nose, tail, and feet. They have medium-sized, droopy ears and originated in the Miami Valley of Ohio in Butler and Warren Counties. This breed stands as the embodiment of perfection in the swine industry.

Name____ Date____________________

7. Breed: ____________________

   These white-bodied pigs have black spots and medium-sized, droopy ears. Part of their ancestry can be traced back to the original Poland China hogs of Warren County, Ohio. This breed has continued to improve in feed efficiency, rate of gain, and carcass quality, as can be proven in testing stations throughout the country. They are popular with farmers and commercial swine producers for their ability to transmit fast-gaining, feed-efficient, meat qualities to their offspring.

8. Breed: ____________________

   This white breed with erect ears is the most recorded breed of swine in the United States and Canada. They are muscular with a high proportion of lean meat and low back fat, in addition to being very sound. They are productive but are more performance oriented and durable than ever. The goal of the breed is to be a source of durable mother lines that can contribute to longevity and carcass merit. The motto "The Mother Breed and a Whole Lot More" indicates improvement and change in the industry.

Breed Bank:

Berkshire	Chester White	Duroc	Hampshire	Tamworth
Landrace	Poland China	Spots	Yorkshire	Meishan

Breed Questions

1. Which breed do you believe is the best? Why?

____

____

____

____

____

2. How does knowledge about swine breeds help the efficiency of a pig farm?

____

____

____

____

____

3. Why are some breeds more popular than others?

____

____

____

____

____

Name_________________________________ Date_______________

4. List any other swine purebred breeds you can think of.

____

____

____

____

____

Name____ Date_____________________

Chapter 20 Matching Activity

Directions: Match the term to the corresponding definition or description.

Terms	Definition
____ 1. swirl	**a.** when the skin has black pigmented spots
____ 2. cryptorchidism	**b.** hair growing in a circular pattern from the roots; usually occurs along the top of the spine
____ 3. freckles	**c.** males with one or both testicles retained
____ 4. indigenous	**d.** nuts and fruits from trees and shrubs
____ 5. heritage breeds	**e.** feeder pig finishers buy feeder pigs and grow them to harvest weight
____ 6. farrow to finish	**f.** a smaller farmer that provides labor and facilities in exchange for a guaranteed price per pig delivered
____ 7. feeder pig	**g.** a model in which the same company that produces the animals also owns the packing plants that slaughter the animals for market
____ 8. contractee	**h.** specific area or country, meaning that they originated from and live in a specific region
____ 9. prolific	**i.** able to produce high numbers of offspring
____ 10. vertical integration	**j.** those that were raised in bygone eras by our ancestors
____ 11. Livestock Conservancy	**k.** pigs from birth to about 10 to 60 pounds
____ 12. mast	**l.** a nonprofit organization that preserves and promotes heritage breeds

Name______________________________ Date______________

Chapter 20 Lab Questions

1. What are the five leading states in hog production? Which of them is number one?

2. Explain how crossbreeding combinations are selected in swine herds.

3. Which new technology has the ability to improve swine breed characteristics?

4. What is your opinion on the ethics of using this technology?

Feel good studio/Shutterstock.com

Chapter 21

Selection and Judging of Swine

Introduction

Most swine operations in the United States use crossbreeding to produce the desired traits of hogs. Terminal crosses are used to prevent the crossbreeding from becoming diluted or losing the traits that the producer wants in the hogs. Modern market hogs are more muscular, produce a larger loin eye area, and have less back fat than those produced only a few years ago. Because swine have relatively short gestation periods and large litters, changing the traits of market hogs does not take a long time.

Quality measurements and carcass grading have a strong influence over the price a hog will bring at market. Ultrasound is a recent technology that can measure the amount of fat-free lean pork in live hogs, helping producers to make decisions about which animals to keep for breeding stock. Pork packers can also use ultrasound measurements to help determine the value of the hog carcass. When swine are evaluated or judged, the evaluator must have knowledge of the desired physical traits for the animal, a knowledge of the anatomy and terms, and the ability to read and review performance data. As a producer, it is necessary to be familiar with parts of the hog. Figure 21-2 in the textbook is a diagram of the pig anatomy with labeled parts.

The four primal cuts of the hog carcass are ham, loin, Boston shoulder (Boston butt), and picnic shoulder. These four cuts represent the most valuable part of the hog carcass. Hog producers try to control the quality of these primal cuts through breeding strategy and by using performance data to increase the probability of hogs with desirable traits.

Performance records are widely used by seed stock producers and market hog producers to evaluate breeding stock. When raising breeding stock, genetic control is an important part of the production program. Performance records are used by both breeders and market hog producers for the genetic improvement of their herds. Much progress has been made in the areas of feed efficiency, growth rate, reproductive efficiency, and carcass quality. This progress has come about because of performance testing and selection programs based on genetic principles, which have resulted in the selection of genetically superior breeding stock. Expected progeny

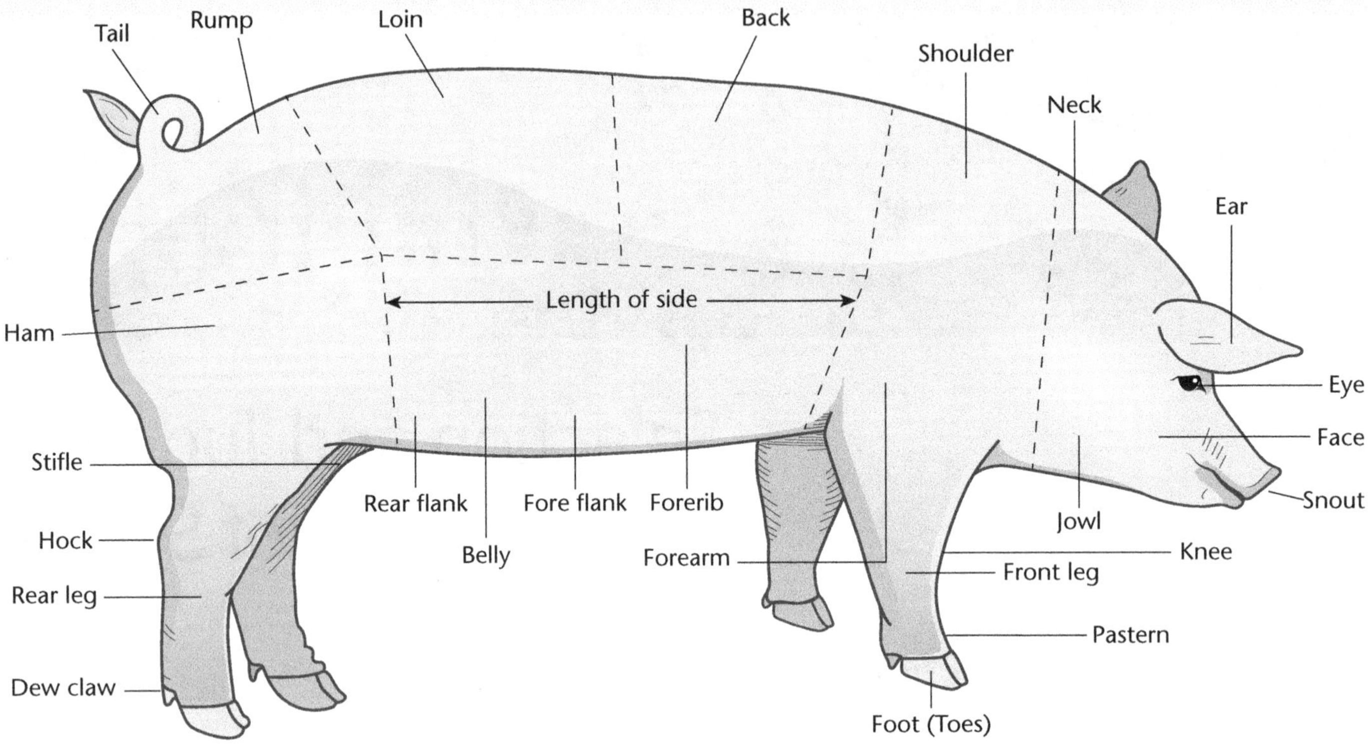

difference (EPD) is a measure of how much of a change in performance on a given trait can be expected from the progeny (offspring) of a breeding animal when compared with the average of all animals in the population. EPDs can be compared within breeds and across herds but not across breeds. Sire summary and EPD values are available from several sources and can be easily accessed online. In addition to performance records, physical criteria are used when selecting breeding stock. These criteria include soundness of the animal and health of the breeding stock herd. When evaluating a group of swine, it is important to use the steps contained in the following list:

1. Understand the class description, situation, or scenario.
2. Set priorities according to class description.
3. Evaluate the performance records.
4. Evaluate visual traits.
5. Decide on final rankings.

These steps can be used when evaluating a group of pigs at a judging contest, at a county fair, when buying a group of hogs, to measure success of a breeding program, or just in herd evaluation and measurement. This chapter provides examples to practice using performance data for evaluation of various groups of swine.

Name______________________________ Date______________

Exercise 21-1 Using Performance Data in Swine Selection

In this exercise, a list of tips, facts, and definitions are given. These will be helpful to evaluate two sample groups of swine using their performance data. The groups will need to be ranked in order of the pigs with the most desirable traits ahead of those with less desirable traits.

Tips and Facts

There are three categories of performance information: REPRODUCTION, GROWTH, and COMPOSITION. Performance records can be presented as (1) individual measurements, (2) ratios and/or indexes, and (3) genetic merit estimates (EPDs).

Category: Usefulness:	Individual Records Good	Ratios & Indexes Better	EPDs Best
REPRODUCTION	#farrowed	Dam's SPI	Born Alive EPD
	#weaned		
	21-day litter wt.	Dam's Index	SPI EPD
	Teat count		
	Dam's record		Maternal Index EPD
GROWTH	Average daily gain	Gain ratios	Days to 230 EPD
	Weight/day of age	NSIF Index	Terminal Index EPD
	Days to 230 lbs	F/G ratios	
	Feed efficiency (F/G)		
COMPOSITION	Back fat scan	BF ratio	BF EPD
	Loin muscle area		Terminal Index EPD

Figure 21-1 Categories of performance information.

Definitions

Sow productivity index (SPI): a measure of milking ability and prolificacy. The index combines number of pigs born alive and 21-day litter weight. It is adjusted for sow parity. The index is calculated with EPDs.

Terminal sire index (TSI): a measure of growth, efficiency, and back fat. The index is calculated with EPDs.

Maternal line index (MLI): places more emphasis on reproductive traits than growth traits; it is used when selecting replacement gilts. The index is calculated with EPDs.

BF: back fat scan.

LEA: loin muscle area.

Name____ Date____________________

EPD number born alive (NBA): predicts the number born alive for daughters' litters relative to their farrowing group average. An individual with an EPD of +0.5 would be expected to produce daughters that would farrow litters with 0.5 more pigs than a sow with NBAEPD 0.0.

EPD litter weight: predicts the 21-day weight for their daughters' litters. An individual with an EPD of +3.5 would be expected to produce daughters that would wean litters 3.5 pounds heavier at 21 days than a sow with an EPD of 0.0.

EPD days to 230 pounds: predicts the growth performance of offspring in days to 230 pounds live weight. An individual with an EPD of 23.0 would be expected to produce progeny that reach 230 pounds 5 days faster than progeny of a parent with an EPD of +2.0.

EPD back fat: predicts the genetic contribution of potential parents for back fat of their progeny at 230 pounds live weight. For example, a boar with a back fat EPD (BFEPD) of −0.04 would be expected to sire progeny 0.04 inch leaner (10th rib BF) at 230 pounds than a boar with BFEPD of 0.00. Also, a gilt with BFEPD −0.05 would produce offspring 0.06 inch leaner than a gilt with BFEPD 0.01.

Directions: For the following examples of swine groups to be evaluated, use the performance data provided to rank the animals in order of most desired performance data to least desired performance data. For each animal, explain why they were ranked in that way using the performance information.

Situation 1: Hampshire Boars

The boars will be pasture bred to Yorkshire/Landrace sows. All offspring will be sold and marketed on a value-based marketing system. Make an initial ranking based on the given situation and the following production data.

The Hampshire boars in this class are going to be used as terminal sires. All offspring are going to be sold on a value-based marketing system. Therefore, the offspring need to be lean and heavily muscled to achieve high market premiums. No replacements are kept, so disregard the maternal data. Structural soundness is not a major priority in this class because the boars will be used on pasture versus confinement. The priorities for this class are leanness, muscle, and growth.

Boar #	Back Fat	Days to 230 lb.	Loin Eye Area	# Born Alive	# Weaned	21-day Wt.
1	0.53	158	7.34	13	10	131
2	0.75	174	6.78	13	10	131
3	0.85	183	6.09	13	10	131
4	0.74	169	6.68	11	7	88

Figure 21-2 Performance data for sampled boars.

Name____ Date____________________

Rank these animals based on the situation and the performance data given for this group:

____

____

____

____

____

____

Explain your reasons for ranking the swine in this order:

____

____

____

____

____

____

Situation 2: Chester White Gilts

Rank the gilts as they should be kept in a seedstock herd that profits mainly from the sale of Chester White × Yorkshire F1 females to owners of terminal crossbreeding programs (terminal boars bred to Chester White × Yorkshire gilts). You and your customers raise all hogs in confinement. Your customers market all progeny on a value-based system.

The Chester White gilts in this class are going to be bred to Yorkshire boars to produce F1 females that are going to be used in a terminal crossbreeding program. The F1 gilts will be bred to terminal sires. All progeny will be marketed on a value-based system. Therefore, the F1 gilts must excel in maternal traits in order to produce large litters of fast-growing market hogs.

Because the F1 female offspring will be expected to raise large litters, a high-quality underline should be emphasized along with the maternal figures. With the F1 gilts being placed in confinement, soundness is a priority. Even though the Chester White gilts will be used for maternal traits, they still must not cause a major setback in growth or leanness. The priorities for this class are maternal excellence, soundness, and high-quality underlines.

#	Days to 230lb EPD	10th Rib Fat EPD (in)	21-day Litter Wt. EPD (lb)	# Born Alive EPD
1	−2.9	+0.04	+0.5	+0.00
2	−2.4	+0.03	+3.7	+0.60
3	+0.8	−0.02	+1.7	+0.25
4	−2.6	+0.03	+3.7	+0.60

Figure 21-3 Performance data for sampled gilts.

Name____ Date__________________

Rank these animals based on the situation and the performance data given for this group:

Explain reasons for ranking the swine in this order:

Name____ Date__________________

Chapter 21 Matching Activity

Directions: Match the term to the corresponding definition or description.

Term

____ 1. contemporary group
____ 2. estimated breeding value (EBV)
____ 3. expected progeny difference (EPD)
____ 4. porcine stress syndrome (PSS)
____ 5. terminal crossbreeding
____ 6. hybrid vigor
____ 7. Boston butt
____ 8. seedstock producer
____ 9. low genetic potential
____ 10. maternal
____ 11. paternal
____ 12. balance
____ 13. underline

Definition

a. a specific cross in which all pigs produced are sold for harvest
b. an inherited neuromuscular disease
c. a group of animals that are similar in a number of characteristics and have been raised under the same management practices
d. heterosis
e. a selection index because it combines information from several sources to determine the genetic merit of the individual; total estimated genetic value of an animal
f. one of the four primal cuts of the hog carcass
g. measures how much difference in performance on a given trait can be expected from the progeny of the breeding animal when compared with the average of all animals in the population
h. refers to the mammary development of the hog; there should be a minimum of six nipples per side, none of which are inverted, scarred, or otherwise nonfunctional
i. all parts of the hog should be properly proportioned
j. producers using genetic improvement programs who keep good records, do performance testing, use selection indexes that relate to important economic characteristics, and follow good selection principles to keep pork competitive with other sources of meat in the marketplace
k. low-quality genetics
l. mother (dam/sow) line genetics
m. father (sire/boar) line genetics

Name____ Date____________________

Chapter 21 Lab Questions

1. Why are quality measurements important in the swine industry?

2. How is ultrasound used in the industry?

3. What are the four most important primal cuts of a pig?

4. What steps should be followed when evaluating a group of swine?

Feel good studio/Shutterstock.com

Feeding and Management of Swine

Introduction

Swine production requires intense management because quality factors affect the price directly. To remain competitive, swine producers must select breeding stock that will produce lean hogs and maintain feed efficiently. Management techniques such as split-sex and phase feeding can help increase production efficiency.

There are two main types of swine enterprises: those that produce purebred animals and commercial hog producers. Purebred animals only make up about 1 percent of the swine population in the United States. They are primarily raised for seedstock genetic improvement; the swine industry is a specialized industry, and the purebred producer must have thorough breeding management practices. The commercial industry has various types of operations such as feeder pig production, feeder finishing, and farrow-to-finish operations.

There are many types of feeds and sources of energy and protein that are used in the swine industry, but the two most common are corn and soybean meal oil. Corn has a high feeding value and soybean has a high level of available protein. Almost all commercial swine are produced in confinement housing for all phases of production, but there are some smaller operations that use pasture management to supply feed and protein while raising hogs. Hogs raised through pasture management have a slower rate of weight gain than confinement hogs. In this system, swine are rotated throughout the available pastures, allowing those fields that are vacant time to rest and regrow. Keep in mind that pigs are not ruminants. Because they lack a rumen and fiber-digesting microbes, pigs are unable to digest fiber well. They cannot survive on plants alone, which is a major reason pasture management is not a large portion of the industry.

Whether fed a grain-based diet or pasture management, hogs have requirements for minerals, vitamins, and a clean, adequate source of water. Salt, ground limestone, and dicalcium phosphate are common sources of sodium, chlorine, calcium, and phosphorus. These minerals are important for many functions of growth and reproduction. Trace minerals and vitamins are added to swine rations by using commercial feeds and mineral and vitamin premixes. Clean, fresh water needs to be readily available at free choice to the animal. Feeding rations differ for each stage of swine development, and a producer must manage the feeding program effectively.

Management of a sow through gestation and farrowing is labor intensive and a fragile time for both the sow and the litter of piglets. There are many strategies that are used to safely manage the sow and raise the piglets. A producer can save more pigs by being present when sows are farrowing, making sure all of the piglets nurse to receive colostrum and that they are clean and dry. There are several necessary procedures that should be carried out to prepare the piglet for growth. Because the piglets' neurological sensors are underdeveloped, they cannot feel as much pain, and they recover quickly. A producer should clip needle teeth, ear-mark pigs, disinfect and clip the navel cord, and dock tails one to three days after farrowing. Iron shots should also be administered two to four days after birth. Male pigs that will not be used for breeding should be castrated at two weeks old. A disease prevention program should be in place for piglets before they are weaned and enter groups with other pigs the same size.

The two exercises in this chapter focus on the various feeds that can be fed to sows during gestation and the volume needed, as well as farrowing practices once gestation has finished.

iStock.com/Baloncici

Name________________________________ Date______________

Exercise 22-1 Feed Impacts on Gestation Activity

Most operations that breed sows and raise them through gestation pay special attention to ensure that the sows get enough energy from feed to properly develop through the stages of pregnancy. Each producer has different available feeds depending on the region and the price of various commodities used in swine feed. In this exercise, there are several situations describing the type of feed available as the base feed for the ration as well as the metabolizable energy needed for the sow. Use the tables in the textbook for reference when developing the feed ration.

Directions: For each situation, determine how much of the base commodity is needed in the ration as well as all other ingredients needed to achieve the desired metabolizable energy levels. Record these values in the feed ration section of each situation.

Situation 1

The producer has good sources of oats, soybeans, and corn. The desired metabolizable energy for the operation is 1441 kcal/lb.

Feed ration: ________________________________

____

____

Situation 2

The producer has a small supply of corn, soybean, and wheat middlings. The desired metabolizable energy for the operation is 1338 kcal/lb.

Feed ration: ________________________________

____

____

Situation 3

The producer has a large supply of hard winter wheat, some soybeans, and dehydrated alfalfa meal. The desired metabolizable energy for the operation is 1352 kcal/lb.

Feed ration: ________________________________

____

____

Name_______________________________________ Date____________________

Exercise 22-2 Farrowing Practices Activity

In this exercise, assume the role of the farrowing manager at a large farrow-to-finish operation. Being an operation with up to 10 sows farrowing a day, there are several employees helping to care for the sows and the piglets. The manager is responsible for successful farrowing, so a standardized operating procedure (SOP) sign is posted in the farrowing building, and an instruction binder is left in the office that outlines farrowing procedures.

Directions: Develop a standardized operating procedure. It should outline all of the considerations and procedures that need to take place during farrowing through weaning. This sheet should explain to employees what the procedures are for care and management of the sow and piglets and why these steps are necessary.

SOP for Farrowing

____

____

____

____

____

____

____

Name______________________________________ Date__________________

Chapter 22 Matching Activity

Directions: Match the term to the corresponding definition or description.

Term	Definition
____ 1. Pork Quality Assurance (PQA Plus)	**a.** contained in cottonseed meal; is toxic to hogs
____ 2. scab	**b.** an animal that consumes a variety of materials, both plants and meat
____ 3. fagopyrin	**c.** a disease that attacks barley crops
____ 4. ergot	**d.** consumes plants
____ 5. binding quality	**e.** a fungus that can infest rye and triticale
____ 6. gossypol	**f.** a management education program with major emphasis on swine herd health and animal well-being
____ 7. flushing	**g.** means to increase the amount of feed fed for a short period of time
____ 8. phase feeding	**h.** refers to how well the feed particles stick together in the pellet
____ 9. omnivore	**i.** a feed additive that increases efficiency in hog production
____ 10. herbivore	**j.** a photosensitizing agent that can cause rashes and itching when white pigs are exposed to sunlight
____ 11. carnivore	**k.** consumes meat
____ 12. feral	**l.** pigs are weaned and isolated in a clean facility for finishing
____ 13. free choice	**m.** feeds are constantly available, and animals consume it as they desire
____ 14. complete mixed feed	**n.** a period of sexual dormancy between two estrus periods
____ 15. restricted feeding	**o.** limited feeding; means feeding 75 to 80 percent of full feed
____ 16. batch farrowing	**p.** wild
____ 17. anestrus	**q.** all ingredients are mixed together in a feed
____ 18. segregated early weaning (SEW)	**r.** pigs are weaned from the sow at 14 to 16 days and placed in an isolated, clean facility that is specifically designed for a nursery operation
____ 19. medicated early weaning (MEW)	**s.** arranging the breeding program so that groups of sows farrow at regular intervals throughout the year
____ 20. tail docking	**t.** cutting off the pig's tail ¼ to ½ inch from the body

Name________________________________ Date________________

Chapter 22 Lab Questions

1. Why are purebred swine important to the rest of the industry?

2. Using Table 22-1 in the textbook, what are the two most common sources of energy and protein in the hog industry?

3. What is the difference between confinement and pasture management in the swine industry?

4. How old should a gilt be when she is first bred and how much should she weigh? How old should a boar be before being used in a breeding program?

5. Several important management practices are performed between farrowing and weaning. List five that should be done.

Feel-good studio/Shutterstock.com

Chapter 23

Diseases and Parasites of Swine

Introduction

Swine production is a challenging industry because hogs are susceptible to many different diseases and parasites. Disease and parasite problems cause millions of dollars in losses each year in the U.S. swine industry.

Herd health problems have become more complex because of certain changes that have taken place in management in recent years. First, more hogs are being raised in confinement. In addition, hogs are being fed rations to push them to maximum growth rates. These two management changes create more stress on the pigs that may lead to an increase in herd health problems. Also, because of faster transportation, diseases may be spread farther in a shorter period of time.

Prevention of health problems is the key to reducing losses. The major factors in maintaining healthy swine are sanitation, selection of healthy breeding stock, and proper management.

There are numerous infectious diseases that can affect swine, so a producer must be knowledgeable of prevention strategies and treatment procedures. Younger pigs are the most susceptible to disease. Vaccinations, sulfa drugs, and antibiotics are useful in the prevention and treatment of many of these diseases. The use of specific pathogen-free (SPF) hogs that have been born and raised in a sterile environment is also helpful in controlling some diseases.

Observing the vital signs in an animal can help in the early detection of health problems. Vital signs will vary according to activity and environmental conditions, but normal vital signs in swine are (1) temperature of 102.0 to 103.6°F, (2) pulse rate of 60 to 80 heartbeats per minute, and (3) respiration rate of 8 to 13 breaths per minute.

Some health problems arise because of feeding factors. Proper feeding levels and maintaining adequate body condition helps to reduce disease. The inclusion of minerals and vitamins is essential in preventing many diseases.

External parasites such as lice and mites can cause many health issues for hogs. A common condition is mange, which is caused by a tiny white or yellow mite that bores into the skin. The skin becomes irritated and results in the animal rubbing the affected areas. Mange is most severe

Figure 23-1 Pseudorabies (also known as Aujeszky's disease or mad itch) is a viral disease in swine, often causing newborn piglets to die.

in the fall, winter, and spring, and it spreads rapidly from hog to hog. Insecticides can be used to control mange, but all safety instructions and withdrawal periods must be followed.

Internal parasites can be very destructive internally and lead to severe sickness and low rates of gain in swine. The most serious internal parasite of hogs is the large roundworm. Hogs that are infested with internal parasites do not grow or gain weight as fast as other hogs, and they are more likely to have other health problems. Signs of a serious infestation of internal parasites are thinness, rough hair coat, weakness, and diarrhea. Some other worms that are major problems are lungworms, nodular worms, whipworms, and intestinal threadworms. A strict program of sanitation and treatment with drugs is necessary to control internal parasites.

Disease organisms may be carried into hog production areas by visitors. It is important to restrict the number of visitors allowed in and out of hog production facilities. Any visitor entering the facility should be wearing disinfected footwear. A container of disinfectant and a brush for disinfecting shoes should be kept at the entrance for the use of persons entering the facilities. There are many steps that can be implemented with a disease prevention program.

The exercise in this chapter will explore some swine diseases in detail. Understanding the causes, signs, and treatment of the diseases is essential for developing a herd health program.

Name______________________________ Date______________

Exercise 23-1 Diseases That Affect Swine

Pigs are susceptible to more diseases than most species of livestock. A producer must understand all possible diseases and implement a prevention plan. When a disease has been contracted, a producer must be able to identify the signs and begin treatment and isolation of infected animals. As a swine producer, explain the protocol for disease identification and control.

Directions: For the diseases listed here, determine (1) the causes of the disease, (2) the signs that a pig has contracted the disease, (3) the effect the disease has on health and production, and (4) the treatment and prevention methods that should be used.

Abscesses

1. ______________________________
2. ______________________________
3. ______________________________
4. ______________________________

Actinobacillus pleuropneumoniae

1. ______________________________
2. ______________________________
3. ______________________________
4. ______________________________

Atrophic rhinitis

1. ______________________________
2. ______________________________
3. ______________________________
4. ______________________________

Name________________________________ Date________________

Avian tuberculosis

1. ________________________________
2. ________________________________
3. ________________________________
4. ________________________________

Brucellosis

1. ________________________________
2. ________________________________
3. ________________________________
4. ________________________________

Cholera

1. ________________________________
2. ________________________________
3. ________________________________
4. ________________________________

Clostridial diarrhea

1. ________________________________
2. ________________________________
3. ________________________________
4. ________________________________

Name______________________________ Date_______________

Swine dysentery

1. ______________________________
2. ______________________________
3. ______________________________
4. ______________________________

Edema

1. ______________________________
2. ______________________________
3. ______________________________
4. ______________________________

Eperythrozoonosis

1. ______________________________
2. ______________________________
3. ______________________________
4. ______________________________

Exudative epidermitis (greasy pig disease)

1. ______________________________
2. ______________________________
3. ______________________________
4. ______________________________

Name____ Date____________________

Chapter 23 Matching Activity

Directions: Match the term to the corresponding definition or description.

Term	Definition
____ 1. abscess	**a.** swellings filled with pus; the result of a bacterial infection that enters the body through the nose or mouth
____ 2. specific pathogen free (SPF)	**b.** located in the snout of the pig; they filter and warm the air before it reaches the lungs
____ 3. rotavirus	**c.** an (after death) examination can accurately identify the cause of an animal's death and help to prevent spread of a disease through the herd
____ 4. turbinate bone	**d.** hogs that come from breeding stock that are surgically removed, generally by cesarean section from the sows under antiseptic conditions; these hogs are raised in disease-free conditions
____ 5. postmortem	**e.** has similar signs as clostridial diarrhea
____ 6. pork checkoff	**f.** transmits harmful organisms to the swine unit; they can include people, cats, birds, dogs, and rats
____ 7. vectors	**g.** program which collects money from producers based on hogs marketed

Name________________________________ Date______________

Chapter 23 Lab Questions

1. What are the normal vital signs for pigs?

2. What are the important factors in preventing losses in a swine herd?

3. Give an example of an internal parasite. How does it affect the hog? How is it treated?

4. Give an example of an external parasite. How does it affect the hog? How is it treated?

Feel good studio/Shutterstock.com

Chapter 24

Swine Housing and Equipment

Introduction

Pigs can be raised on pasture, but most swine raised in the United States are raised in confinement housing. Housing is needed for farrowing, growing, and finishing hogs. A small number of producers still use pasture for all or part of the hog production cycle. These producers usually sell their pork as value added.

Swine operations must be very strategic in their location for many reasons. The location of the farm in relation to neighbors and others is important because there are often complaints of the smell and traffic of daily operation at a hog farm. The direction of prevailing winds and the location of residences in the area must be considered when designing and building a swine facility.

Many producers in the swine industry are investing large amounts of capital into buildings and equipment to efficiently produce pork products. Modern confinement housing is equipped with proper ventilation to control temperature and moisture. Farrowing houses may be equipped with a heating source depending on the climate where the farm is located. Slotted floors are common to hog farms. They are an added cost, but they increase efficiency in manure removal and help keep hog pens clean. The plan for a hog enterprise must include fencing, handling equipment, feeding and watering equipment, crop production equipment, and any other automated equipment used on the farm.

Some farms with intensive management systems have higher equipment costs. There are many options for facility use from sanitation rooms for employees equipped with showers and sanitation suits, as well as the use of video cameras to monitor the animals especially during farrowing. Exercise 24-1 will give an example of equipment used for one swine operation. Study the data and determine the total market value of the farm's equipment.

Figure 24-1 Example of modern swine housing.

Name____ Date____________________

Exercise 24-1 Equipment Inventory Worksheet

All equipment used by a farm or ranch should be included in an equipment resources inventory. The inventory should include any important information such as the model or serial numbers found on the equipment as well as the size, age, and condition of each piece of equipment. It should also indicate whether the equipment is owned, leased, or borrowed. Finally, a current value should be included for each piece of equipment. Determining the value of equipment can be done by determining the book value of the piece of equipment and then deducting the depreciation from the value. The depreciation schedule will provide a value of the equipment that is equal to the original cost minus any accumulated depreciation.

Equipment Example

Directions: Use the following example of equipment used on a small hog farm for equipment information and values. Enter available information into the Equipment Resources Inventory Worksheet to determine the total book value and market value (with depreciation) of all equipment on the farm. You may also add additional examples, as directed by your teacher. Note that under Condition, G = good, F = fair, and P = poor. Under Ownership, O = owns, L = leased, and B = borrowed.

This operation has an older line of equipment and does not invest as much as some producers because their land and housing payments are quite high. The farm sees a depreciation value of 5 percent of the book value on all its equipment. All equipment is owned outright by the farm with no financing loans, and most was purchased when the farm started five years ago, unless stated otherwise. The equipment is all in good condition because the owners take care of it. The operation owns a 200-horsepower tractor that is 5 years old. It was bought for $79,000. There are 20-foot chisel plows and 15-foot offset plows worth $7,500 each. The farm has several grain drills to cover all the land. All drills are 20 feet; one cost $12,000, one cost $13,000, and the third cost $9,000. There are two farm trucks: One is 5 years old and worth $35,000; the other is only 1 year old and worth $55,000. The farm also has various pieces of livestock equipment ranging from farrowing crates to automatic waters: the total value of this equipment is $8,000. Other miscellaneous ranch equipment is valued at $5,000 in total.

Name________________________________ Date______________

Equipment Resources Inventory Worksheet

List all Equipment Resources of the Operation

(Use Additional Pages if Necessary)

Equipment Name	Model #	Size	Purchase Year	Age	Condition			Ownership			Book Value	Market Value
					G	F	P	O	L	B		

Name____ Date____________________

Chapter 24 Matching Activity

Directions: Match the term to the corresponding definition or description.

Term	Definition
____ 1. warm confinement house	**a.** used by counties, cities, and other local governments to define how property may be used
____ 2. cold confinement house	**b.** maintain desired temperatures regardless of the outside temperature
____ 3. growing period	**c.** a low-cost housing alternative for swine that originated in Canada; has been gaining popularity in the Midwest
____ 4. zoning laws	**d.** time from weaning to about 100 pounds
____ 5. hoop structure	**e.** have temperatures that are only slightly warmer (3 to 10 degrees) than the outside temperature
____ 6. quonset	**f.** these laws required farmers to fence in their crops to protect them from livestock that were allowed to roam freely
____ 7. gestation crate	**g.** open pens used during farrowing to make it possible for the piglets to keep a safe distance away from the sow and avoid being crushed
____ 8. farrowing crate	**h.** semi-circular build with no straight walls
____ 9. free access stalls	**i.** stalls as the primary housing for gestating sows; in this system, sows stay in crates 24 hours a day from the time they are bred until they are moved into the farrowing crates a few days before farrowing
____ 10. finishing facility	**j.** open pens of 10 to 25 sows
____ 11. stock and fence laws	**k.** houses used to finish pigs; they can include open front, modified open front, or totally enclosed facilities

Name____ Date____________________

Chapter 24 Lab Questions

1. What factors should be considered when planning a swine facility? Prepare a list of must-have items.

2. Why is the ventilation system important in a swine facility?

3. List and explain the four major phases of swine production.

feel good studio/Shutterstock.com

Chapter 25

Marketing Swine

Introduction

Marketing swine is an essential part of the business, and the decision of which market to use can make or break the profit margins for a hog producer. On any given day, markets can see variation in the price offered, the services given, and the costs charged to the seller and buyer. These factors change based on the number of hogs at the market that day as well as supply and demand. A producer must carefully study market trends to receive the best returns to the business.

There are four main avenues to market swine: direct marketing, auction markets, terminal markets, and group marketing systems. Most hogs in the United States are marketed through direct marketing; fewer producers market hogs through terminal and auction markets. The trend toward group marketing systems has been growing to help producers get higher prices for their pork.

Most swine in the United States are sold based on the weight of the animal. Producers that raise pork for quality standards get a better price by selling on yield and grade bases. When being marketed, swine are classified according to use, sex, weight, and quality. Butcher hogs and feeder pigs are graded according to USDA official grading standards. The grades of market hogs are U.S. No. 1, U.S. No. 2, U.S. No. 3, U.S. No. 4, and U.S. Utility, a system in which No. 1 indicates the most desirable and Utility is used for the least desirable. This grading system is based on the quality of the lean meat and the percentage of the four lean cuts that the carcass will produce.

Currently, the best prices are typically paid for hogs weighing 220 to 260 lb (99.8 to 117.9 kg). Most producers aim to send hogs to market at 200 to 220 lb (90.7 to 99.8 kg). Large, meaty hogs can be raised over the common weight if the cost-to-price ratio remains positive. Hog prices tend to fluctuate with the seasons; they are generally lower in early spring and higher in the summer. Producers must account for the loss of weight due to shrinkage while the hogs are being transported through the marketing system.

The futures market may be used to reduce the risk of loss by hedging a fixed price. A producer using the futures market should understand the futures market and consult with experts.

iStock.com/Miguel Perfectti

iStock.com/DS70

Name______________________________________ Date____________________

Exercise 25-1 Marketing Methods Lab

A swine producer must choose a market based on the best options for its operation and the option giving the most value per pound of pork. A choice of market depends on many factors such as price, size of animal, convenience for producer, reasons for sale, and several herd considerations.

Directions: For each of the following market descriptions, determine if it is a terminal market, an auction market, group marketing, or direct selling. Be specific as to the method of marketing used.

1. A meat-packing corporation owns the local country buying station in an area of large swine production. The producer has a short distance to transport the swine, thereby decreasing shrinkage. The hogs are bought from the producer at a price based on the fat-free lean percentage of the carcass either on a percentage of the selling price or a fixed fee.

   Marketing method: ____

2. A large hog marketing cooperative is located in several states and increases the bargaining power of individual hog producers. This cooperative negotiates contracts with packers to supply large numbers of hogs from producers that are a part of the cooperative.

   Marketing method: ____

3. The producer consigns swine to a commission firm. There are usually several buyers competing for hogs, which may produce a better price. A commission fee will be taken out of the producer's check received for the hogs.

   Marketing method: ____

4. A hog farmer in a rural area raises slaughter hogs and then has them processed at a private USDA butcher. The producer pays a fee for processing, and all of the pork is returned processed, packaged, and labeled. The farmer then brings the pork to the city using a refrigerated truck and sells the product to restaurants and through farmers' markets.

   Marketing method: ____

5. A local livestock auction barn uses cell phones and video auctions to permit buyers who are not physically present to bid on the hogs.

   Marketing method: ____

Name____________________ Date__________

Chapter 25 Matching Activity

Directions: Match the term to the corresponding definition or description.

Term	Definition
____ 1. boar taint	**a.** major causes of the odor problem with boar meat
____ 2. skatole/androsterone	**b.** when using the futures market, this is done to reduce the risk of loss if prices go down by locking in a price
____ 3. shrinkage	**c.** odor given off by the meat of a boar
____ 4. futures contract	**d.** loss of weight as hogs are shipped to market
____ 5. hedging	**e.** establishes a price for live hogs that are to be delivered at some future date
____ 6. net return	**f.** an increased demand for pork and pork products resulting from checkoff programs from the swine producers return after costs
____ 7. direct marking	**g.** groups negotiate contracts with packers to supply hogs
____ 8. terminal markets	**h.** commission firms can help the producer select the best time to market hogs
____ 9. auction markets	**i.** a male that has not been castrated
____ 10. group marketing	**j.** a male that was castrated while young
____ 11. marketing cooperative	**k.** a young female that has not farrowed and is not showing any signs of pregnancy
____ 12. slaughter swine	**l.** have been established to help pork producers get higher prices
____ 13. feeder swine	**m.** to obtain more buyers for the hogs offered; for example, cell phones and video auctions permit buyers who are not physically present to bid on the hogs
____ 14. barrow	**n.** swine that are killed and sold as meat
____ 15. gilt	**o.** involves selling to packing plants, order buyers, or country buying stations
____ 16. sow	**p.** a male that was castrated after reaching maturity or is beginning to show the physical characteristics of a boar
____ 17. boar	**q.** swine that are sold to be fed to higher weights before harvest
____ 18. stag	**r.** an older female that has farrowed or is showing signs of pregnancy
____ 19. xenotransplantation	**s.** the transplantation of animal tissues or organs into a human

Name____ Date__________________

Chapter 25 Lab Questions

1. What is the pork checkoff program? Describe how it benefits the swine industry.

2. Explain how shrinkage affects market hogs.

3. Describe the different grades a swine carcass can receive. How does this affect price?

4. Explain how the four swine markets work: direct, terminal, auction, and group markets. Why would a hog producer choose to use these markets?

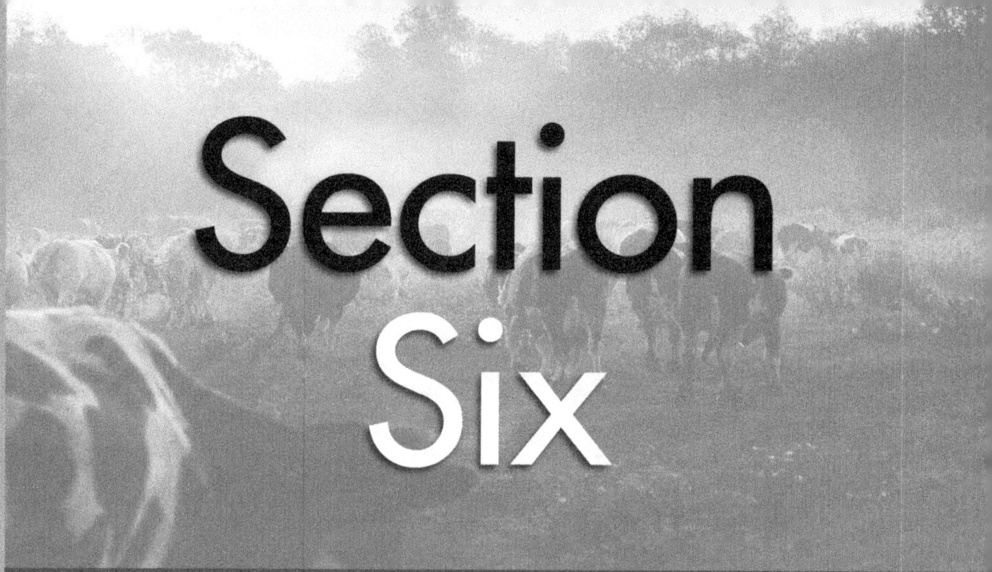

Section Six

Sheep and Goats

Feel good studio/Shutterstock.com

Feel good studio/Shutterstock.com

Chapter 26

Breeds and Selection of Sheep

Introduction

There are more than 40 breeds of sheep in the United States. Each breed has unique characteristics. Some sheep are large, and some are smaller; some have black legs and heads, while others are white. Sheep are a major farm enterprise in the western range area of the United States, but in the rest of the country, farm flocks tend to be a secondary enterprise. In this case, cattle are usually the primary enterprise while sheep are mainly used to help maintain pasture.

Sheep are classified into several categories based on their wool production, meat production, breeding qualities, dairy production, and a combination of other functions. One way to classify sheep is by the type of wool they produce. Breeds of sheep that produce wool are separated into the following categories: fine wool, long wool, crossbred wool, medium wool, carpet wool, fur, and hair sheep. Fine wool breeding is most common in the western range areas, while the medium wool and crossbred wool breeds are more common in the farm flock area of the country. Hair sheep are becoming more popular in all parts of the United States as a low-maintenance meat breed.

As with most species of livestock, breed selection depends on the preference of the producer; therefore, sheep are divided into categories for evaluation. Depending on the function of the breed and the goals of the operation, different traits are valued when selecting sheep. The basic traits for sheep evaluation are overall health, soundness, fleece, conformation, and age. It is necessary to handle sheep when selecting or judging to determine traits such as muscling and finish. Market lambs are judged on the additional trait of finish, which can be measured by touch. Breeding sheep are also judged on condition (finish in market lambs), size, feet, legs, bone, breed and sex character, and fleece.

iStock.com/CanY71

The two exercises in the chapter will refine breed identification skills as well as market lamb and breeding sheep evaluation skills. Knowledge and experience in breed and quality selection are important for all sheep producers.

Name_______________________________________ Date____________________

Exercise 26-1 Sheep Breed Identification

Sheep breeds can be grouped based upon their primary use in the sheep industry. Some sheep are raised to produce meat, and others are raised to produce wool. Some breeds are "ewe" breeds because they make good mothers, and some are "ram" or "meat" breeds because they grow fast. Some sheep are dairy breeds and are raised to produce milk for making cheese. Sheep breeds can also be grouped based on the type of fiber they produce. Some sheep produce fine, high-quality wool, while other breeds produce medium- or coarse-fiber wool. Some breeds have hair and do not need shearing like the wool breeds. Some breeds are classed as "dual purpose" because they are acceptable for either meat or wool production.

Directions: For the following eight classifications of sheep, identify one example of a breed from each classification. Give a detailed description of the breed (history, functions, production characteristics, etc.).

1. Meat breed: ____

2. Fine wool breed: ____

3. Coarse wool breed: ____

4. Hair breed: ____

5. Ewe breed: ____

6. Black-faced breed: ____

Name______________________________ Date______________

7. Dairy breed: ____

____

____

____

8. Dual-purpose breed: ____

____

____

____

Name______________________________ Date______________

Exercise 26-2 Market or Breeding Lamb Lab

Market lambs are judged on the basis of type, muscling, finish, carcass merit, yield, quality, balance, style, soundness, and smoothness. Breeding sheep are also judged on condition (finish in market lambs), size, feet, legs, bone, breed and sex character, and fleece.

During this exercise, the instructor will provide a sample group of show-quality market or breeding lambs to be evaluated. A trip to a local sheep farm may be possible for a sample group, or a video of a judging class of market or breeding lambs may be used.

Directions: Use the following scorecard to describe and evaluate the sample market or breeding lambs. Record the information for four lambs, and then rank the lambs in order of most desirable traits.

	Lamb 1	**Lamb 2**	**Lamb 3**	**Lamb 4**	
Type					
Muscling					
Finish					
Carcass Merit					
Yield					
Quality					
Balance					
Style					
Soundness					
Smoothness					
Rank Group					

Name____ Date____________________

Chapter 26 Matching Activity

Directions: Match the term to the corresponding definition or description.

Term	Definition
____ 1. banding instinct	**a.** ewes produced anywhere in the United States other than the West
____ 2. band	**b.** computerized, performance-based program for selection of sheep
____ 3. staple	**c.** consists of the skin and fleece
____ 4. NSIP	**d.** strong instinct to flock together
____ 5. western ewe	**e.** select body traits of sheep
____ 6. native ewe	**f.** a group or a flock
____ 7. type	**g.** ewes that are produced in the western range area
____ 8. pelt	**h.** refers to the fibers of wool
____ 9. piebald	**i.** having irregular patches of two colors

Name____________________ Date__________

Chapter 26 Lab Questions

1. Describe one breed of fine wool sheep.

2. Describe one breed that is dual purpose.

3. What qualities are ewes evaluated for?

4. What qualities are market lambs evaluated for?

Feel good studio/Shutterstock.com

Chapter 27

Feeding, Management, and Housing of Sheep

Introduction

As in other livestock industries, sheep production can be divided into two categories: purebred and commercial production. There are four common types of sheep management systems based on the time of lambing: (1) fall lambs, (2) early spring lambs, (3) late spring lambs, and (4) accelerated lambing. Purebred producers produce breeding stock for commercial flocks and for other purebred flocks. The major goal of a purebred producer is breed improvement, so keeping good production records is necessary. Commercial producers maintain sheep flocks to produce meat and wool, so they use various breeds and crossbreeds to produce the desired traits.

When managing a flock of sheep, the feeding needs of the animals change as they enter different stages of development. The major stages of development that require specific feeding management are gestation feeding, lactation feeding, feeding the ram, flushing, feeding lambs to weaning, feeding lambs from weaning to market, feeding orphan lambs, and feeding replacement ewes. Each stage requires a custom ration with different amounts of roughage and concentrates. Pasture and good-quality legume or grass-legume hay are used as common roughage sources. Corn, grain sorghum, oats, barley, and wheat are grains that are also commonly used in sheep rations. If good-quality pasture or hay is available, protein supplements are not needed. If necessary, soybean oil meal, cottonseed meal, linseed meal, and peanut meal are all good sources of protein to add to a sheep ration.

It is important that sheep producers do not allow breeding sheep to become too fat or too thin; they need a balanced body condition to effectively reproduce. Many breeds of sheep breed in the fall, but some sheep breed in other seasons as well. In order to have good breeding rates, a producer must prepare the ewe for breeding by clipping the wool from the vulva, udder, inside of the thigh, and around the eyes. The ram should be checked for fertility before the breeding season so there is time to find a replacement sire if needed.

Lamb care is very delicate, and it is easy to have lamb loss if proper management is not used. A producer can save lambs by making sure they nurse and do not become chilled. Depending on the location of the farm, warm confinement facilities may be needed. Lambs need to be docked, castrated, marked, and vaccinated to give them the best chance for survival. More concentrate than roughage is used to feed lambs to market weights.

In this chapter, the exercises provide situations requiring feed management and breeding decisions. Use the principles of sheep management described in the textbook to best complete the exercises.

Vito Giaccari/Pexels

Name____ Date____________________

Exercise 27-1 Feeding Considerations

Maintaining feed with the proper available energy for the requirements of sheep at different stages of growth is imperative for efficient management. Producers try to avoid having sheep become too thin or gain too much condition. The appropriate feeding ration is based on the stage of development of each animal.

Directions: For each of the following sheep at different stages of life, describe the suggested feeding strategy for proper body condition maintenance. (Suggested feeding rations are provided in the textbook.)

1. **Gestation feeding**

____

____

____

____

2. **Lactation feeding**

____

____

____

____

3. **Feeding the ram**

____

____

____

____

4. **Flushing**

____

____

____

____

5. **Feeding lambs to weaning**

____

____

____

____

Name__________________________________ Date________________

6. **Feeding lambs from weaning to market**

7. **Feeding orphan lambs**

8. **Feeding replacement ewes**

Name____ Date____________________

Exercise 27-2 Crossbreeding Activity

Many sheep producers use crossbreeding to grow ewe and ram lambs with desirable traits. Most farms use a three-breed cross for the best results. Although most breeds have similarities, there are traits that are more dominant in one breed over another. The breeds in a cross must be chosen carefully, and the producer needs to have a good understanding of the traits each breed provides. Use Chapter 26 in the textbook as a reference for the traits of each breed.

Directions: Below are five examples of crossbred lambs that possess specific genetic traits. For each example, identify three sheep breeds that can be crossbred to produce a lamb with those traits.

Lamb 1: Breed A____________________ Breed B____________________ Breed C____________________

Description: The traits of this crossbreed include wrinkled skin and medium, angular bodies. They are white faced with wool on their legs. They do not have a strong flocking instinct.

Lamb 2: Breed A____________________ Breed B____________________ Breed C____________________

Description: The ears, nose, face, and legs are white. This crossbreed produces a medium-coarse fleece that averages 4 to 5 lb. They are large in size and have a blocky body type.

Lamb 3: Breed A____________________ Breed B____________________ Breed C____________________

Description: They have no wool on the face and legs. The face, ears, and legs are gray to brown, and the lamb crop is often 150 percent.

Lamb 4: Breed A____________________ Breed B____________________ Breed C____________________

Description: They have gentle disposition and are economical to feed and maintain. The lambs grow rapidly and deposit fat at a slower rate than other breeds. The breed is polled, and the color of the face is reddish brown to bright tan.

Lamb 5: Breed A____________________ Breed B____________________ Breed C____________________

Description: This sheep is polled, and it has a tuft of wool on the forehead. It produces 12 to 16 lb of fleece and is better adapted to wet, marshy areas than are other long wool breeds of sheep.

Name____ Date____________________

Chapter 27 Matching Activity

Directions: Match the term to the corresponding definition or description.

Term	Definition
____ 1. hothouse lamb	**a.** occurs when the uterus protrudes from the vulva
____ 2. accelerated lambing	**b.** refers to shearing around the udder, between the legs, and around the dock
____ 3. drench	**c.** a system that produces three lamb crops in two years
____ 4. tagging	**d.** a male lamb that has been castrated before reaching sexual maturity
____ 5. wether	**e.** a portable gate that can be lengthened or shortened as needed to crowd sheep into a corner
____ 6. prolapsed uterus	**f.** lambs that are sold at 50 to 90 days of age; they weigh 35 to 60 lb
____ 7. extension hurdle	**g.** a large dose of medicine mixed with liquid and put down the throat of the animal
____ 8. enterotoxemia	**h.** a young sheep
____ 9. ewe	**i.** a mature female sheep
____ 10. lamb	**j.** an uncastrated male sheep
____ 11. ram	**k.** (overeating disease) when starting on full feed to protect lambs on rations high in concentrates
____ 12. flock	**l.** a group of animals herded together

Name______________________________ Date______________

Chapter 27 Lab Questions

1. What are the main goals of a purebred sheep producer?

____

____

____

____

____

2. What procedures are essential to keep lambs alive after they are born?

____

____

____

____

____

3. What are the common ingredients used in sheep feeding rations?

____

____

____

____

____

Feel good studio/Shutterstock.com

Chapter 28

Breeds, Selection, Feeding, and Management of Goats

Introduction

There has been a national trend toward higher consumption of goat products, including meat, milk, fiber, and other products. At one time, goat products were only available in health-food stores or specialty shops, but now goat meat, milk, cheese, and butter can be found in local supermarkets across the country. This increase in demand has changed the scope of the goat industry from a small niche market to a large, popular market with many new producers growing goats.

One of the major reasons for the increased production of goats is the growth of the ethnic populations in the United States who view goat meat as customary in their diet. Other reasons for the increase in the production include the ability of goats to utilize land unsuitable for other livestock, an increase in demand for goat products due to heath concerns, the rising popularity of goat meat as delicacies in more upscale restaurants, and the increase in their popularity as show animals.

The three major classifications of goats are dairy, fiber, and meat. There are breeds of goats that were also established for pets/companions and for goatskin. Small herds of dairy goats are common in all parts of the United States due to the increase in hobby farms.

The major breeds of dairy goats found in the United States are Alpine, LaMancha, Nigerian Dwarf, Nubian, Oberhasli, Saanen/Sable, and Toggenburg. Dairy goats produce much less milk than cattle, but they are still selected for similar traits and high milk production. High-producing dairy goats average three to four quarts of milk per day.

Although there are many breeds of fiber sheep, the three fiber goat breeds that are common in the United States are Angora, Cashmere, and Miniature Silky Fainting Goats. Fiber goats are evaluated on the basis of body type and fleece. Breeding animals are further selected on the basis of age and fertility.

The common meat type goat breeds in the United States are Boer, Kiko, Kinder, Myotonic, Pygmy, Savanna, and Spanish. When selecting goats for breeding stock, producers must consider the purpose, functionality, and durability of the animal.

Goats are ruminants and have digestive systems similar to those of cattle, so roughage may be used as a large part of the ration when feeding goats. Producers will supplement grain when low-quality roughages are being fed in order to provide additional nutrients during growing and lactating periods.

Each goat operation will have different housing needs depending on the size and scope of the operation. More equipment and facilities are necessary for dairy goats. These facilities depend on the number of goats owned and the convenience desired by the producer. When running a dairy goat operation, an essential aspect of the business is keeping the milk and milking equipment clean.

This chapter explores a few of the numerous goat breeds in existence throughout the United States. Complete the exercises in order to develop knowledge and skill in breed identification and goat facility design.

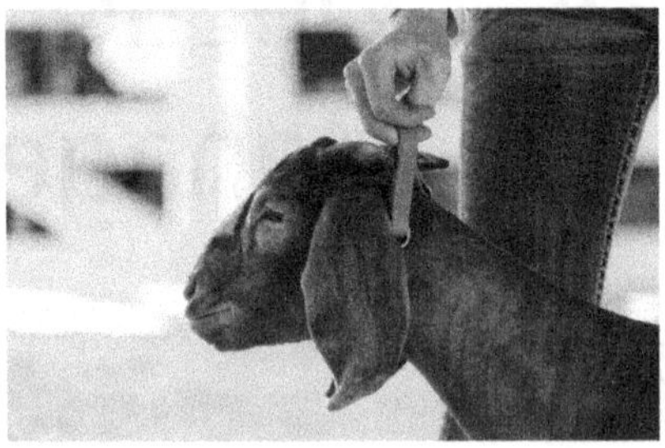

iStock.com/Candice Estep

Name____ Date__________________

Exercise 28-1 Goat Breed Identification

There are more than 90 breeds of goats available to producers worldwide depending on what desired traits suit the livestock operation. Knowledge of the breed descriptions is necessary when selecting a breed to produce.

Directions: For each number, use the description to correctly identify the breed name.

1. ________________
2. ________________
3. ________________
4. ________________
5. ________________
6. ________________
7. ________________
8. ________________

Descriptions

1. This breed originated in France, has upright ears, and can be any color or combination of colors. It has a straight face, has medium to short hair, and is medium to large in size.
2. This is the only breed developed in the United States. It has either "gopher" or "elf ears." Any color or combination of colors is acceptable, and it has short, fine, glossy hair.
3. A Swiss breed of rugged bone, this breed is medium to large in size and either white or cream in color. It has short and fine hair, is erect-eared, and has either a straight or dished face.
4. A Swiss breed known for upright ears, straight faces, and chamois color, it has a black belly and a light gray to black udder. One of the smaller Swiss breeds, it is a minimum of 28 in. in height and is the newest recognized breed by the ADGA.
5. This breed was originated in the Himalayan Mountains of Asia and has a straight or concave nose, pendulous ears, and twisted horns. It is usually a small, white breed with a long, fine, and lustrous mohair fiber coat. The fine underwool is a valuable product called cashmere. This breed is known primarily as a browsing animal.
6. This breed came from West and Central Africa and the Caribbean. Dwarf, short legged, hardy, and alert, its profile should have a dished appearance with a broad, strong, and well-muscled jaw. It has a small compact body, and its main colors are white caramel, caramel, gray agouti, black agouti, and charcoal.
7. This breed originated in India and Egypt and is known for its high quality, high butterfat concentration, and high milk productivity. It has a strong, convex facial profile between the ears and the muzzle and long, bell-shaped, wide ears. It can have any color pattern and has short, glossy, and fine hair.
8. Of Swiss origin, this breed is medium in size, has upright ears and a dished or straight face, and is solid-colored, varying from light fawn to dark chocolate. It has white ears with dark spots in the middle, two white stripes down the face from each eye to the muzzle, white hind legs, and a white triangle on either side of the tail. It is known for its high milk productivity.

Name______________________________ Date______________

Exercise 28-2 Design a Goat Facilities Lab

Designing a dairy goat facility takes many different factors into consideration. Some of these considerations include using loose pens or tie-stall barns, design of the building structure, the size of pens and room for each animal, providing water, providing feed, climate considerations, corrals and handling equipment, and various styles of milking facilities and equipment, as well as milk storage equipment. This exercise focuses on the design of a dairy facility that includes all of the necessary equipment.

Directions: The following facility design is an example of a basic dairy goat operation. Fill in the facility map with all of the equipment necessary to milk goats. The added equipment and facilities should be drawn on the map and labeled. Creating a key to the drawings of equipment may help keep the map neat and legible.

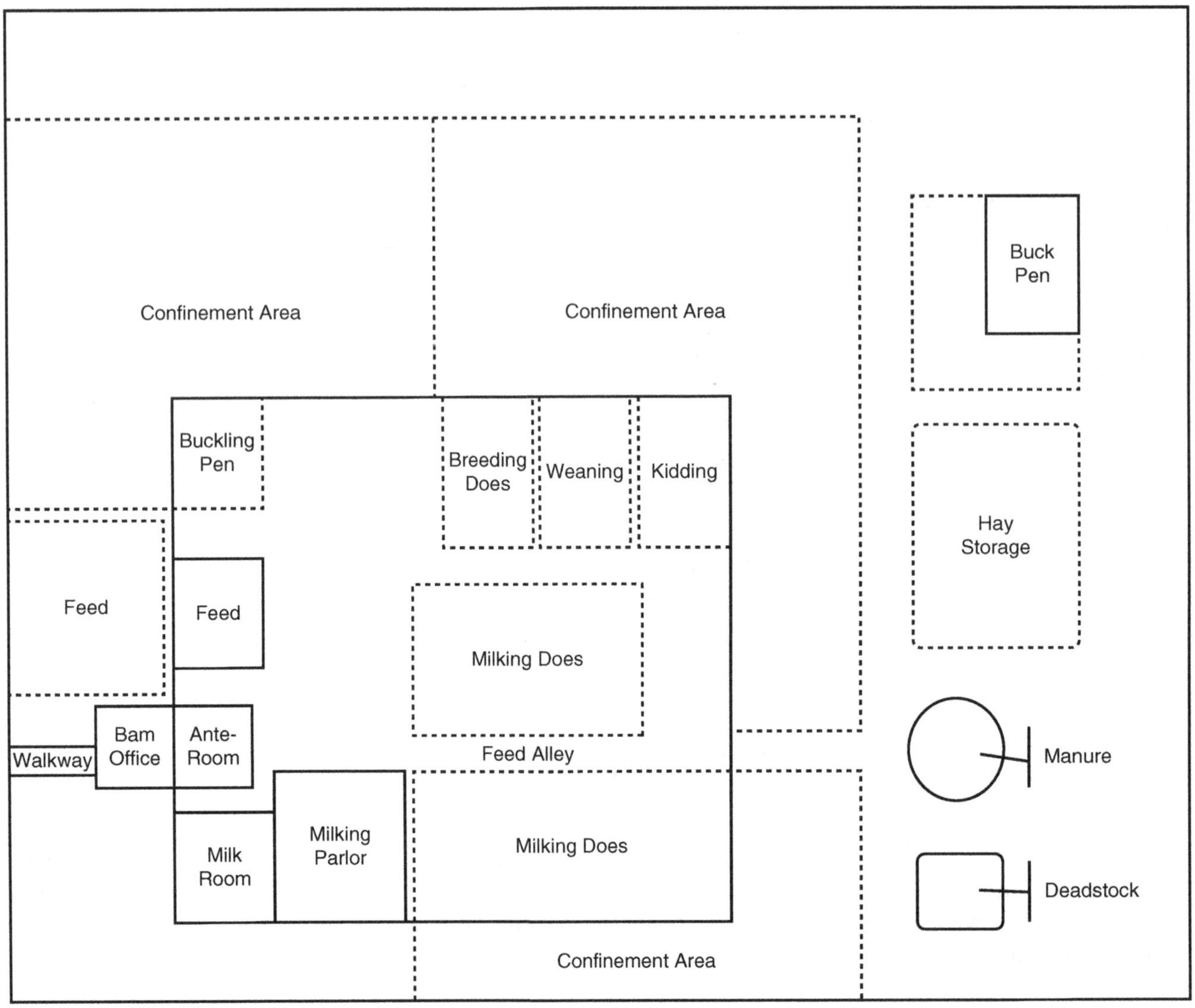

Name______________________________ Date______________

Chapter 28 Matching Activity

Directions: Match the term to the corresponding definition or description.

Terms	Definitions
____ 1. buck	**a.** young goat under one year of age, either sex
____ 2. doe	**b.** refers to spasm or temporary rigidity of the muscles
____ 3. doeling	**c.** a small segment of a larger market
____ 4. kid	**d.** male goat, any age
____ 5. wether	**e.** the shoots, twigs, and leaves of brush plants
____ 6. niche market	**f.** a projection of skin hanging from the chin
____ 7. browse	**g.** female goat, any age
____ 8. wattle	**h.** refers to the soft down or winter undercoat of fiber produced by most breeds of goats
____ 9. cashmere	**i.** an unbred female goat
____ 10. myotonic	**j.** male goat castrated when young
____ 11. animal husbandry	**k.** to protect the goats from the weather after shearing in the spring
____ 12. yearling	**l.** providing for the health, nutritional needs, housing, and general care of their animals to be successful
____ 13. dual purpose	**m.** winter undercoat of fiber produced by most breeds of goats, except the Angora
____ 14. mohair	**n.** goat over one year and under two years of age, either sex
____ 15. down	**o.** mineral deposits on the equipment
____ 16. kemp	**p.** the fleece of the Angora goat
____ 17. polyestrous	**q.** having more than one period of estrus in a year
____ 18. disbudding	**r.** known for milk and meat production
____ 19. milkstone	**s.** another term for dehorning
____ 20. caping	**t.** fibers are large, chalky white hairs

Name______________________________________ Date____________________

Chapter 28 Lab Questions

1. Why are goat populations in the United States increasing in popularity? Explain.

2. Which factors must be considered when selecting goats for breeding purposes? Why are they important to each breeder?

3. What are the other terms used to describe myotonic goats?

Feel good studio/Shutterstock.com

Chapter 29

Diseases and Parasites of Sheep and Goats

Introduction

In order to maintain a good health program for sheep and goats, a producer must focus on prevention of infection rather than treatment. Most treatments are costly and not very effective. When sheep and goats become sick, they rarely recover because, by the time signs of illness are noticeable, the infection has usually progressed beyond treatment. Without a proper treatment plan, profits on sheep and goat farms decrease significantly. Good management, feeding, vaccination, and sanitation programs help prevent health problems. The involvement of a veterinarian is recommended when developing a treatment plan.

Sheep and goats have many diseases in common with cattle, but there are several that specifically affect these two species. These include blue tongue, enterotoxemia, foot rot, foot abscess, dysentery, pneumonia, scrapie, tetanus, and several others. The symptoms of many of these diseases are similar, so identifying the sickness may take laboratory tests and the assistance of a veterinarian. Many of these illnesses can be prevented using available vaccines. If an animal becomes infected, treatment with antibiotics, drenching, and isolating the sick animal are basic control measures.

The most common internal worm parasites of sheep and goats are stomach worms, intestinal worms, lungworms, and liver flukes. The eggs of most of these internal parasites are spread over pasture grasses through animal feces. Frequent pasture rotation and management are needed to prevent or reduce infections. Allowing pastures to rest between grazing periods gives time for larvae to die before they are consumed while animals are grazing.

Raising sheep and goats can be challenging because a large number of nutritional health problems can affect the animal. Many of these infections can be prevented or cured with good feeding management. Some health problems are related to nutritional factors, including night blindness, milk fever, impaction, pregnancy toxemia, and consuming poisonous plants.

The exercises in this chapter involve recommending tapeworm prevention strategies and identifying various parasites that affect sheep and goats.

Name______________________________ Date______________

Exercise 29-1 Tapeworm Lab

The infection of worms presents a significant production threat to sheep, costing the industry millions of pounds a year. The main worms affecting sheep are roundworms, tapeworms (cestodes), and lungworms. Figure 29-1 shows the process that takes place when a sheep is infected with tapeworms. A producer must understand the cycle of parasite infection and change management strategies to reduce or eliminate the infection of parasites such as tapeworms.

Directions: Study the path that a tapeworm takes as in infects a sheep. Once the cycle is understood, suggest five management strategies that could help a sheep or goat farmer reduce or eliminate the presence of tapeworms in the animals.

Tapeworm Prevention Recommendation:

1. ______________________________
2. ______________________________
3. ______________________________
4. ______________________________
5. ______________________________

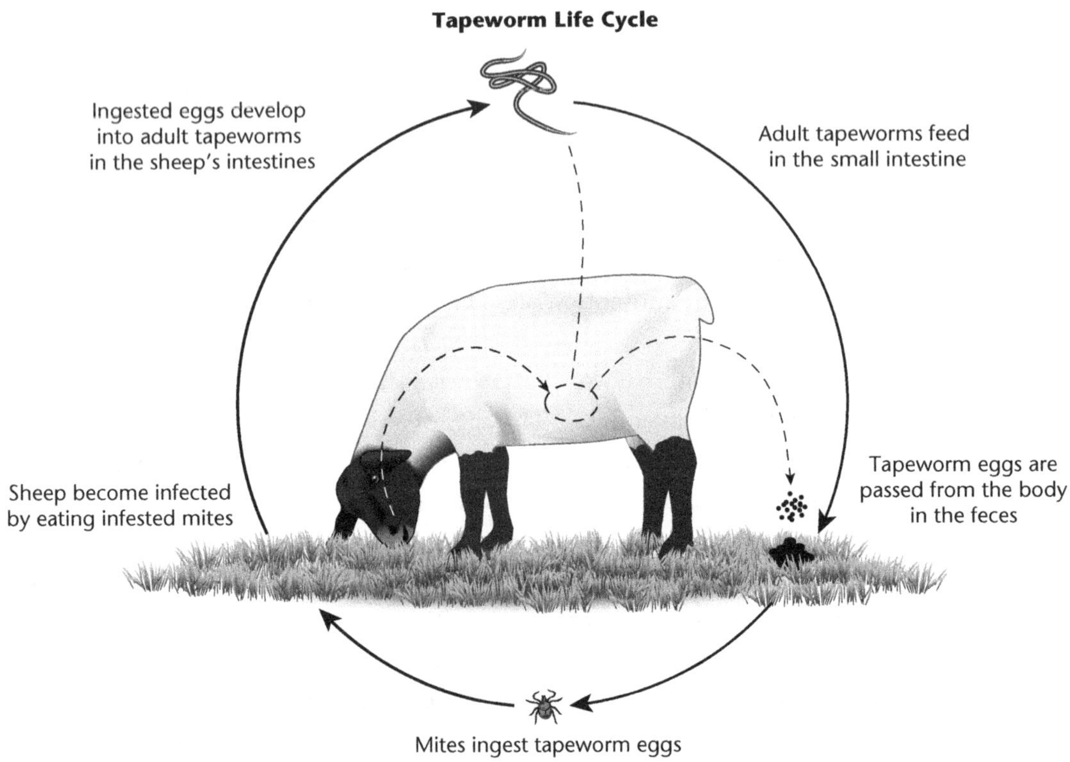

Figure 29-1 Life history of a tapeworm of sheep.

Name________________________________ Date____________________

Exercise 29-2 Parasite Identification Lab

Directions: For each of the following internal parasites of sheep and goats, determine (1) the scientific name, (2) the characteristics of the parasite, (3) the effect the parasite has on the animal, and (4) possible treatment or prevention strategies. Some information may need to be gathered using internet sources.

Common Stomach Worm:

1. ________________________________
2. ________________________________
3. ________________________________
4. ________________________________

Bankrupt Worm:

1. ________________________________
2. ________________________________
3. ________________________________
4. ________________________________

Lung Worm:

1. ________________________________
2. ________________________________
3. ________________________________
4. ________________________________

Liver Fluke:

1. ________________________________
2. ________________________________
3. ________________________________
4. ________________________________

Name____ Date__________________

Chapter 29 Matching Activity

Directions: Match the term to the corresponding definition or description.

Term	Definition
____ 1. virus	**a.** involuntary muscle contractions
____ 2. toxemia	**b.** a self-reproducing agent that is considerably smaller than a bacterium and can multiply only within the living cells of a suitable host
____ 3. convulsion	**c.** a biological pathogen that causes a disease; generally bacterial, fungal, or viral
____ 4. causative agent	**d.** a condition resulting from the spread of bacterial toxins through the bloodstream
____ 5. encephalitis	**e.** chemical compounds used for deworming animals
____ 6. bacteria	**f.** single-celled microorganisms that can cause infection and disease in animals and humans
____ 7. zoonosis	**g.** an infectious disease that is transmitted between animals and humans
____ 8. anthelmintic	**h.** an anthelmintic that is administered orally
____ 9. drench	**i.** an inflammation of the brain

Name______________________________ Date______________

Chapter 29 Lab Questions

1. Why is the treatment of sickness not usually successful with sheep and goats?

2. What is the most effective strategy to control infection of disease and parasites?

3. List the most common parasites in sheep and goats.

4. Describe the life cycle of parasites and how they enter the bodies of the animals.

Feel good studio/Shutterstock.com

Chapter 30

Marketing Sheep, Goats, Wool, and Mohair

Introduction

The sheep and goat industries in the United States are small in comparison with other livestock industries such as swine, poultry, beef, and dairy. Although the numbers of sheep marketed in the United States have been decreasing, the industry is looking to reverse the trend through niche markets and the growth of small start-up operations.

The American Lamb Board and American Wool Trust provide national promotion of sheep and wool products through advertising, public relations, and retail promotions. There are several marketing methods available to sheep producers: terminal markets, local pools, sale barns, direct to the packer, or electronic marketing. The producer must choose the option that works best for the operation based on the kinds of markets available locally, prices paid, the numbers of lambs to be sold, and the transportation options.

Sheep are classed according to age, use, sex, and grade. Age classes are 1) lamb, 2) yearling, and 3) sheep. The use classes are 1) slaughter sheep, 2) slaughter lambs, 3) feeder sheep, 4) feeder lambs, 5) breeding sheep, and 6) shearer lambs. The sex classes are divided into 1) ewes, 2) rams, and 3) wethers (castrated males). When marketing meat sheep, grades are given for live animals that are directly related to the quality and yield grades of the carcasses they will produce. Yield grades are based on 1) the measure of back fat, 2) predicted yield of the carcass, 3) quality grades on the body conformation, and 4) amount of finish.

Wool is marketed through local buyers, wool pools, cooperatives, warehouse operators, or by direct sale to wool mills. Each producer will choose a wool market depending on several local factors, including number of fleeces to market, transportation requirements, price paid, and convenience. The value of the fleece is determined by the amount of clean wool that it produces, the length of the fibers, the density, and the diameter of the wool fibers.

Goat meat is called chevon and is mostly marketed directly to the customer. Young kids that weigh 30 to 40 lb are marketed as cabrito, which is considered a delicacy. Explore marketing opportunities of sheep through the exercises in this chapter.

iStock.com/Osarieme Eweka

Figure 30-1 Retail wool products.

Name________________________________ Date____________________

Exercise 30-1 Classing and Grading of Sheep

Sheep are classified into four categories: age, use, sex, and grade. These factors change the value of the sheep carcass, the texture, and the taste. Live grades of sheep are based on quality and estimated yield. This exercise will provide information on two different sheep to be classed and graded.

Directions: Based on the information provided in the following examples, determine (1) the class, (2) the quality grade, and (3) the yield grade for each sheep.

Sheep 1

This ram is almost three months old and weighs 55 lb. It is moderately wide and thick in relation to its length; it has moderately plump and full legs, a moderately wide and thick back, and moderately thick and full shoulders. The lean flesh and external finish are firm; the flanks are moderately full and firm. It has a back-fat measurement of 0.16 to 0.25 in.

Class: ____________________

Quality grade: ____________________

Yield grade: ____________________

Sheep 2

This ewe is nine months old. It is moderately narrow in relation to its length; has slightly thin, tapering legs; and has slightly narrow and thin back and shoulders. It has moderately firm, lean flesh and external finish and has slightly full and firm flanks. It has a back-fat measurement of 0.26 to 0.35 in.

Class: ____________________

Quality grade: ____________________

Yield grade: ____________________

Name____ Date____________________

Exercise 30-2 Marketing Wool

There are four common methods to market sheep wool. A producer must understand each market option and choose the best one according to convenience and the price received. Part 1 will require a knowledge of each marketing method, and Part 2 will examine sheep operations in the local area and determine which marketing method best suits the local wool business.

Part 1

Directions: Describe each of the following marketing options in detail. Explain who is involved, how money is exchanged, and what the producer can expect when it takes the wool to the market.

Wool pools: ____

____

____

____

Cooperatives: ____

____

____

____

Warehouse operators: ____

____

____

____

Direct sale to wool mills: ____

____

____

____

Part 2

Evaluate the local wool industry. What size sheep farms are in the county? Which wool markets already exist in a reasonable distance? Suggest an existing local market where producers can take their wool. Explain why this is the best marketing choice.

Suggested market for local producers: ____

____

____

____

Name____ Date____________________

Chapter 30 Matching Activity

Directions: Match the term to the corresponding definition or description.

Term	Definition
____ 1. breeding sheep	**a.** the number of hanks of yarn that can be spun from 1 lb of wool top
____ 2. shearer lamb	**b.** process that uses longer fiber wools
____ 3. cutability	**c.** intermingling of fat with lean
____ 4. feathering	**d.** the wool or mohair produced by a single shearing
____ 5. grease	**e.** western ewes that are sold back to farms and ranches to be bred to produce more lambs
____ 6. spinning count	**f.** used in the making of clothing
____ 7. wool top	**g.** refers to animals not finished enough for slaughter; they are shorn and fed to a higher level of finish before slaughter
____ 8. apparel wool	**h.** the presence of impurities in the fleece
____ 9. worsted	**i.** refers to the yield of closely trimmed, boneless retail cuts that come from the major wholesale cuts
____ 10. clip	**j.** partially processed wool
____ 11. mutton	**k.** the term used when marketing processed meat from goats
____ 12. lamb	**l.** the meat from sheep over a year old
____ 13. hank	**m.** yarn that is 560 yards in length
____ 14. woolen	**n.** young kid goats that weigh 30 to 40 pounds and are considered a delicacy and are especially popular for barbecuing
____ 15. chevon	**o.** animals younger than one year
____ 16. cabrito	**p.** making a weaker yarn

Name______________________________ Date_______________

Chapter 30 Lab Questions

1. Which two associations provide promotion of sheep products?

2. What is the most common method of marketing chevon?

3. What qualities are wool grades based upon?

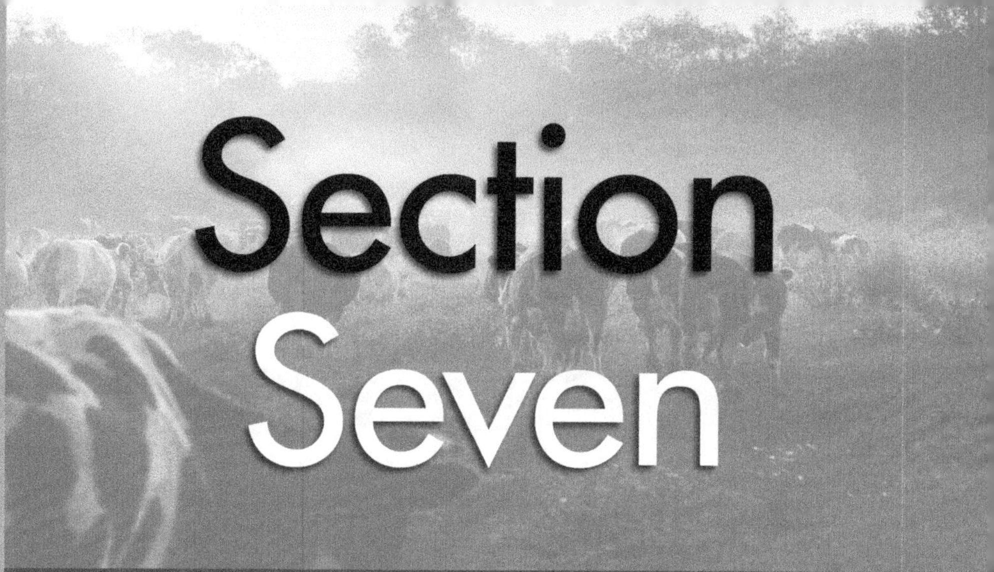

Section Seven

Horses

Feel good studio/Shutterstock.com

Selection and Judging of Horses

Introduction

The use of horses in the United States has changed drastically over the past 150 years. Horses are no longer needed as a tool for agriculture, transportation, and military use. The horse population has decreased over this time, but horses have remained popular and have been adapted for use to fulfill many other purposes. Today, it is estimated that the total horse population of the United States is about 7.25 million. Of these, 75 percent are used for personal recreation and pleasure; the remaining 25 percent of horses are used for ranching, racing, breeding, and commercial riding.

Although the population of horses has decreased, the horse industry is still big business. In the United States, the equine industry is an important contributor to the economy with more than 4.6 million people involved in the industry as owners, service providers, employees, and volunteers. Much of this economic activity is generated by the horse racing industry and by recreational uses of horses.

There are many breeds of horses, and each breed has been developed to perform specific tasks. Horses are classified as either a light breed, a pony breed, or a draft breed. Most of the horses in the United States are of the light breeds. Today, there are fewer breeds of draft horses in the United States due to the decreased use of draft horses in the agriculture industry. Ponies are smaller than light horses. Draft horses are larger than light horses.

Horses should be selected based on conformation, intended uses, the age of the animal, sex, and soundness. Breed selection is a matter of function and personal preference. The five basic colors of horses are bay, black, brown, chestnut, and white. There are different variations of color, which include dun, gray, palomino, pinto, and roan. Horses may suffer from several unsoundnesses and blemishes. Unsoundnesses affect the horse's physical performance. The most serious unsoundnesses affect the horse's feet and legs. A blemish is visible but does not affect the horse's performance.

Name____ Date____________________

Exercise 31-1 Breed Selection

The most popular breeds of horses in the United States, based on purebred registrations, are Quarter Horse, Thoroughbred, Paint, Appaloosa, and Arabian; these make up 70 percent of the total horse population. Many other breeds exist in small numbers. This exercise will test your ability to identify various breeds of horses.

Directions: For each of the following pictures of horses, identify the breed and list three breed characteristics (A, B, and C).

1.

Courtesy of Sari Levin

Figure 31-1

Breed: ______________________________

A. ______________________________

B. ______________________________

C. ______________________________

2.

Courtesy American Paint Horse Association © 2007

Figure 31-2

Breed: ______________________________

A. ______________________________

B. ______________________________

C. ______________________________

Name________________________________ Date______________

3. 

Courtesy of the Arabian Horse Association

Figure 31-3

Breed: ______________________________

A. ______________________________

B. ______________________________

C. ______________________________

4. 

Courtesy of Jillian Paige Photography/ Appaloosa Horse Club, Inc.

Figure 31-4

Breed: ______________________________

A. ______________________________

B. ______________________________

C. ______________________________

5. 

Courtesy of *The American Quarter Horse Journal*

Figure 31-5

Breed: ______________________________

A. ______________________________

B. ______________________________

C. ______________________________

Name______________________________ Date______________

6.

Courtesy of Vesty/Tennessee Walking Horse Breeders and Exhibitors Association

Figure 31-6

Breed: ____

A. ____

B. ____

C. ____

7.

Courtesy of Pony of the Americas Club Inc.

Figure 31-7

Breed: ____

A. ____

B. ____

C. ____

Name____ Date____________________

Exercise 31-2 Anatomy Identification Exercise

Knowledge of the basic exterior equine anatomy is important for anyone working with horses.

Directions: Label the following figure of a horse. Identify the part of anatomy at each line.

Name________________________________ Date________________

Exercise 31-2 Unsoundness Activity

The following table lists several unsound characteristics of horses. These blemishes can be aesthetic or impede the horse's performance and can even keep them from being registered.

Table 31-1: Unsoundnesses and Their Causes

Unsoundness/Stable Vice	Area Affected	Symptoms and Causes
Heaving	Lungs	- Loss of elasticity in the lungs - Contraction of the flank muscles during expiration - Can be caused by a dusty atmosphere or hay
Bowed Tendons	Limbs	- Thickening of the back surface of the leg immediately above the fetlock
Bog Spavin	Limbs	- Soft fluctuating enlargement located at the upper part of the hock due to a distention of the joint capsule
Bucked Knee	Limbs	- Knee is in front of the plumb line
Camped Out	Limbs	- Knee is behind the plumb line
Contracted Heel	Limbs	- Back of the foot becomes narrower than normal
Founder/Laminitis	Limbs	- Inflammation of the sensitive laminae that attach the hoof to the fleshy portion of the foot - Happens when horses consume unlimited grain
Navicular Disease	Limbs	- Inflammation of the navicular bone and bursa
Wind Sucking		- Pressing the upper front teeth on an object, while pulling backward and sucking air into the stomach
Cribbing		- Horses grasp an object between their teeth and apply pressure, gradually gnawing away at the object
Weaving		- A rhythmical shifting of the weight from one front foot to the other - Caused by an idleness in confined quarters
Roaring	Lungs	- Paralysis of the vocal cord nerve resulting in roaring or whistling while inhaling
Blindness	Head	- Impaired vision, cloudiness of the cornea, and white coloring
Bad Mouth	Head	- Describes various jaw or tooth misalignments
Poll Evil	Head	- A lesion or sore on the poll of the horse
Quidding	Head	- The horse drops food while eating due to mouth or teeth problems
Fistulous Wither	Body	- Inflammation affecting the withers (similar to poll evil)
Sweeney	Body	- Muscle atrophy of the shoulder
Knocked Down Hip	Body	- A fracture of the external angle of the hip bone

Name____ Date____________________

Activity: Using the table, create an entertaining game that teaches these blemishes to the people playing the game. This game will be played by the class, and each student will facilitate the game he or she creates.

Game Suggestions (alternative games may be created if approved by the instructor)

Pictionary: Create pictures for 10 of the unsound characteristics. For each picture, there should be an information page behind it that shows the correct name of the blemish and the description. (This could be done by hand on flash cards or using a computer program and projection.)

Jeopardy: Create a Jeopardy!-style game using the unsound characteristics. There should be at least four categories, with five questions in each category. This game can be created on the computer. There are several Jeopardy!-style templates that can be located online. A handmade version of the game is also acceptable.

Wheel of Unsoundness: Create a spinning wheel with at least 15 unsound characteristics. The players will spin the wheel and answer with the description of whichever characteristic they land on. When making the game, make a page with the description of each blemish from the wheel. This key will be used to tell players if they are right or wrong. This can be made by hand or by using a template found online.

Name____ Date____________________

Chapter 31 Matching Activity

Directions: Match the term to the corresponding definition or description.

Term	Definition
____ 1. light horse	**a.** young horse of either sex up to one year of age
____ 2. hand	**b.** male ass
____ 3. tobiano	**c.** common name for the ass; the ass is smaller than the horse, has longer ears, and has a short, erect mane
____ 4. overo	**d.** the offspring when a jack is crossed with a mare (female horse)
____ 5. donkey	**e.** used mainly for riding, driving, and racing; measures 14–2 to 17 hands at the withers
____ 6. jack	**f.** unit to measure horse height; each unit equals four inches
____ 7. jennet	**g.** type of paint horse whose head is marked in the same way as that of a solid-colored horse; the legs are white, at least below the knees and hocks; there are regular spots on the body
____ 8. mule	**h.** female less than three years of age
____ 9. foal	**i.** female ass
____ 10. filly	**j.** paint horse with variable-color head markings; the white usually does not cross the back between the withers and the tail; one or more legs are dark colored, and the body markings are irregular and scattered
____ 11. colt	**k.** male less than three years of age
____ 12. mare	**l.** bad habits in horses; the most common are cribbing, wind sucking, halter pulling, and kicking
____ 13. stallion	**m.** the ways in which a horse moves; examples are walk, trot, gallop, canter, stepping pace, running walk, fox-trot, rack, pace, and amble
____ 14. gelding	**n.** mature male four years of age or older
____ 15. gait	**o.** defect that affects the usefulness of the horse
____ 16. equine	**p.** mature female four years of age or older
____ 17. color breed	**q.** members of the horse family
____ 18. unsoundness	**r.** male that has been castrated
____ 19. blemish	**s.** imperfection that does not affect the usefulness of the horse
____ 20. vice	**t.** refers to groups of horses whose registration is based primarily on coat color

Name______________________________________ Date__________________

Chapter 31 Lab Questions

1. Describe the current uses of horses in the United States.

2. What is the economic impact of the horse industry in the United States?

3. Which unsound characteristics are most serious? Why?

Chapter 32

Feeding, Management, Housing, and Tack

Introduction

Horses are fed diets that contain a large percentage of roughage. Horses have a simple stomach rather than a rumen, so a horse owner must understand the digestive characteristics needed in horse feed. The amount of roughage and concentrate that a horse needs varies tremendously depending on the size, stage of growth, condition, and amount of work the horse needs to perform. Mature, idle horses that do not use a lot of energy can be fed on a ration composed solely of roughage. Grass and legume pastures can provide much of the roughage a horse needs. For horses that are working or pregnant, or for growing foals, some types of concentrates are needed in the ration to provide adequate energy and protein.

Grains such as corn, milo, barley, and wheat can be used in a horse ration, but oats are the preferred grain. Soybean meal can be used as a supplement for protein, but supplements are not often needed. Many small horse operations purchase commercial protein supplements or complete pelleted rations. Three major concentrates that need to be part of a horse ration are salt, calcium, and phosphorus. Regular feeding and watering are important to managing a horse's health.

A proper breeding program is essential in many equine operations. Most mares are bred in the spring because they are more likely to conceive a foal. Pregnant mares need a proper feeding ration, exercise, and extra attention as they approach foaling. Mares should be monitored during foaling in case there is a need for assistance. Proper feeding and training of the foal are essential. Weaning at four to six months is a common practice.

All equine operations should be equipped with proper equipment for grooming, trimming hooves, saddling, and tacking up horses. Tack (equipment) is expensive and should be kept clean and in good repair. The Western and English saddles are the two most common types in use. Bridles and bits are selected based on how the horse is going to be used.

This chapter requires the student to identify proper feed rations for horses at different stages of development. It also includes a hands-on horse tack lab.

Name____ Date____________________

Exercise 32-1 Feeding for Appropriate Size and Age

When feeding horses, having a separate ration for animals at different stages of development is important. The energy and protein requirements for each stage vary, and it is important that a horse is not fed too much or too little of any nutrient. Use information from the textbook chapter as a reference for this exercise.

Directions: For each of the following horse descriptions, suggest an appropriate ration that could be fed to each horse and how much to feed. The ingredients of the ration should be listed as a weight (lb) to be fed (daily allowance).

1. **A weanling colt weighing 600 lb**

   Ingredients: ____

   Daily allowance: ____

2. **A yearling weighing 900 lb**

   Ingredients: ____

   Daily allowance: ____

3. **A 1000-lb working horse being fed a conditioning ration**

   Ingredients: ____

   Daily allowance: ____

4. **A mare that is eight months pregnant**

   Ingredients: ____

Name_______________________________________ Date__________________

Daily allowance: ____
____
____

5. **A 1000-lb two-year-old mare (not bred)**

Ingredients: ____
____
____

Daily allowance: ____
____
____

6. **A 1000-lb lactating mare**

Ingredients: ____
____
____

Daily allowance: ____
____
____

7. **A 1000-lb 22-year-old idle horse**

Ingredients: ____
____
____

Daily allowance: ____
____
____

8. **A 300-lb suckling foal**

Ingredients: ____
____
____

Daily allowance: ____
____
____

Name____ Date____________________

Exercise 32-2 Horse Tack Identification

Horse equipment (tack) is very expensive. The ability to identify and maintain horse tack is just as important as the ability to properly place and adjust the equipment on a horse. Part 1 is an identification activity, and Part 2 will require hands-on work tacking a horse.

Part 1

Directions: Label each arrow on Figures 32-1 and 32-2. Tear the diagrams out and turn them in to the instructor for grading.

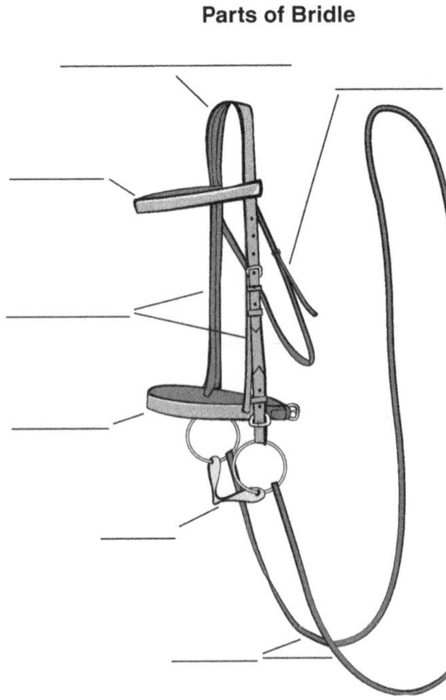

Figure 32-1 Diagram of a bridle.

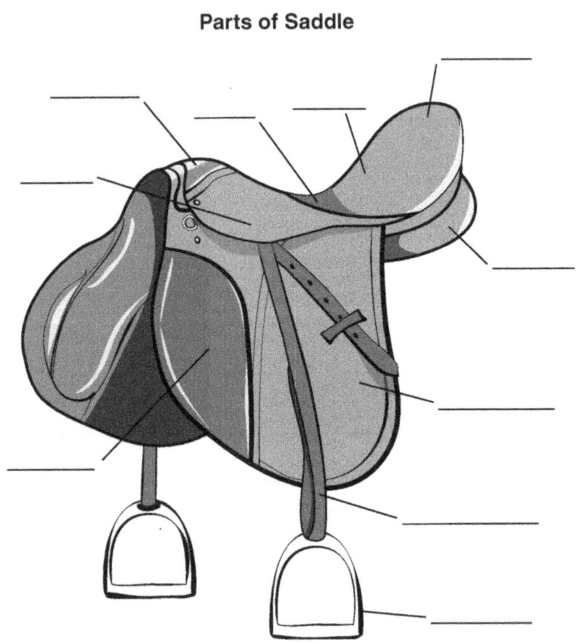

Figure 32-2 English saddle diagram.

Name______________________________ Date______________

Part 2

Directions: Perform the following tacking tasks.

Materials Needed: Demonstration horses, several styles of tack equipment (grooming equipment, saddles, girths, bridles, leg gear, and leather straps for tying demonstrations).

Task 1

Tying Skills: Using sample leather straps, practice each of the knots from Figure 32-3. Show each knot to the instructor for approval before moving on to the next style of knot.

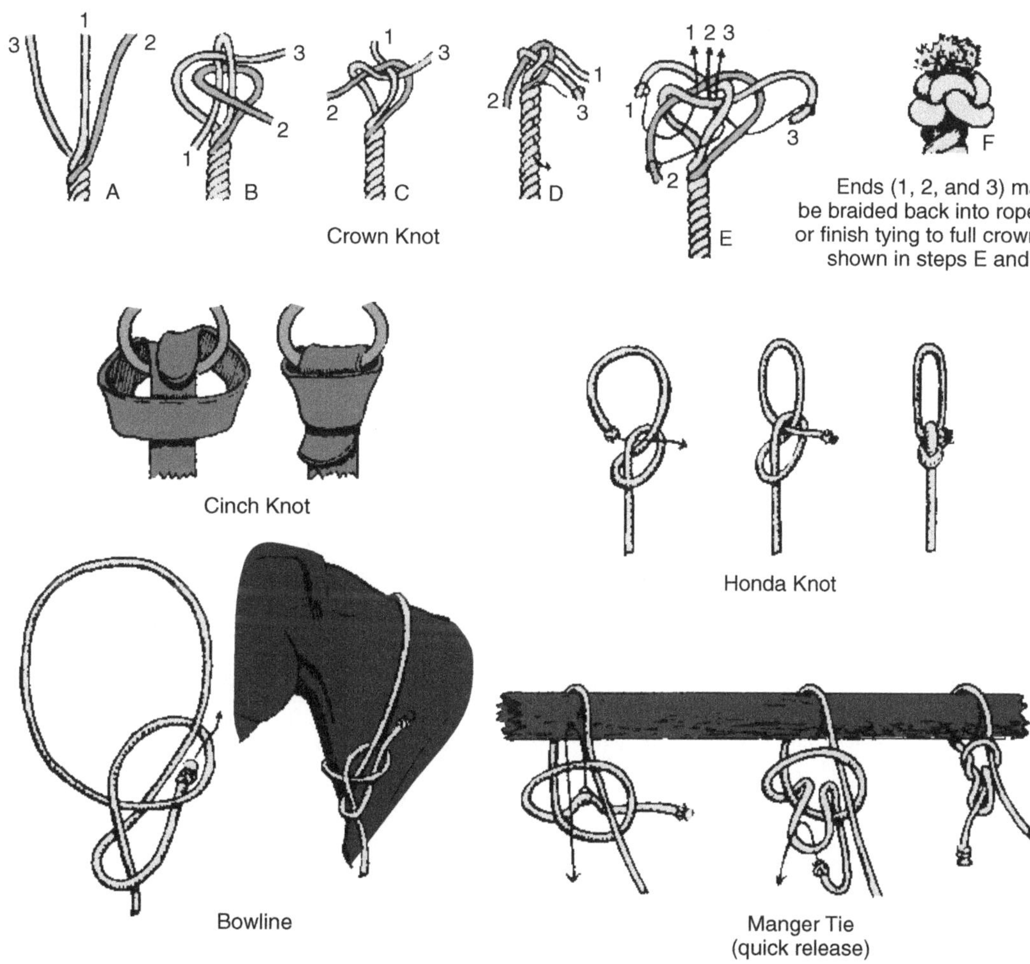

Figure 32-3 Knots commonly used by horse owners.

Name____ Date__________________

Task 2

Grooming: Tie the horse's lead to a rail and begin grooming practices. Curry comb, brush, and pick hooves with pick. Once the instructor indicates that the horse is properly groomed, move on to the next task.

Task 3

Saddling: Prepare the saddle to mount on the horse; lift one stirrup and place the saddle on the horse in the proper location. Adjust all girths and stirrups to properly fit the horse and rider. Once the instructor indicates that the horse is properly saddled, move on to the next task.

Task 4

Bridle Adjustment: Choose the proper bridle and bit for the horse. Place the bit in the horse's mouth and the bridle over the muzzle and ears. Adjust the bridle so that it is not too tight or too loose. Once the instructor has approved the bridle, the horse is ready to ride.

Name________________________________ Date_______________

Chapter 32 Matching Activity

Directions: Match the term to the corresponding definition or description.

Term	Definition
____ 1. meconium	**a.** used for tying or leading horses
____ 2. imprinting	**b.** person who works on horses' feet
____ 3. roached mane	**c.** the feces that are impacted in the bowels during prenatal growth
____ 4. float	**d.** the gear used to attach a horse or other draft animal to a load
____ 5. farrier	**e.** its purpose is to hold the bit in the horse's mouth
____ 6. bridle	**f.** long-handled rasp with a guard to prevent injury to the horse during treatment
____ 7. tack	**g.** piece of hard material put in the horse's mouth to help control the horse
____ 8. bit	**h.** behavior-shaping process that takes place within the first 24 hours of a foal's life
____ 9. halters	**i.** equipment used for riding, handling, caring for, and showing horses
____ 10. harness	**j.** is trimmed down to the crest of the neck, leaving the foretop and a wisp of mane at the withers
____ 11. martingale	**k.** device that prevents the horse from lifting its head too high

Name_______________________________ Date_______________

Chapter 32 Lab Questions

1. Explain how a horse's diet is different from other livestock discussed in this lab book.

2. How do the seasons affect horse reproduction?

3. Make a list of tack that should be kept when owning a horse for performance.

Feel good studio/Shutterstock.com

Chapter 33

Diseases and Parasites of Horses

Introduction

Parasites and equine disorders can have costly impacts on horses and horse owners. The symptoms of these health issues can cause weight loss and loss in performance, and can result in costly treatment methods; in the worst case, the animal may die. Proper feeding and good management help reduce losses from diseases and parasites. Cleanliness and sanitation of the equine facilities and pastures are the most important aspects of a disease and parasite prevention program. Proper exercise and grooming are also necessary to help keep horses in good health and free from parasites and bacteria.

There are vaccines available for many types of equine diseases, but there are some for which no vaccine exists. The most serious diseases of horses are distemper, encephalomyelitis, equine infectious anemia, and equine influenza. When developing a vaccination program, a veterinarian familiar with the horse should be consulted. Knowledge of the diseases and symptoms is necessary for early detection. The longer the disease goes untreated, the lower the chances of successful treatment.

External and internal parasites pose a risk to horses every year. There are many different parasites that use horses as a host. The common external parasites of horses are flies, lice, mites, ringworms, and ticks. These parasites can cause digestion problems, skin irritation and swelling, and constant discomfort to the horse. To control external parasites, there are several brands of insecticides that can be applied to eliminate the parasite.

More than 150 different species of internal parasites affect horses. The most serious internal parasites are strongyles, ascarids, pinworms, bots, and tapeworms. Heavy infestations of internal parasites lead to poor physical condition and even death of the horse. While horses of all ages are affected, young horses are more susceptible and seriously affected. For internal parasites, the most effective control is taking prevention measures based on the life cycle of the parasite.

The following exercises will develop knowledge of the various equine disorders and the treatment/prevention strategies available and provide a parasite control lab and worksheets.

Name______________________________________ Date____________________

Exercise 33-1 Equine Disorders

Directions: For each of the following equine disorders, describe (1) the symptoms of the disorder, (2) prevention of the disorder, and (3) the treatment when symptoms are presented.

Anhydrosis

1.____

2.____

3.____

Anthrax

1.____

2.____

3.____

Azoturia (Monday morning sickness)

1.____

2.____

3.____

Colic

1.____

2.____

3.____

Cushing's syndrome

1.____

2.____

3.____

Distemper (strangles)

1.____

2.____

3.____

Name______________________________ Date______________

Exercise 33-2 Parasite Life Cycle Lab

Figure 33-1 presents an example of a horse parasite's life cycle. Understanding the life cycle is important to treat and prevent the infection of many types of parasites. In this exercise, complete the parasite identification activity, as well as the treatment and prevention charts.

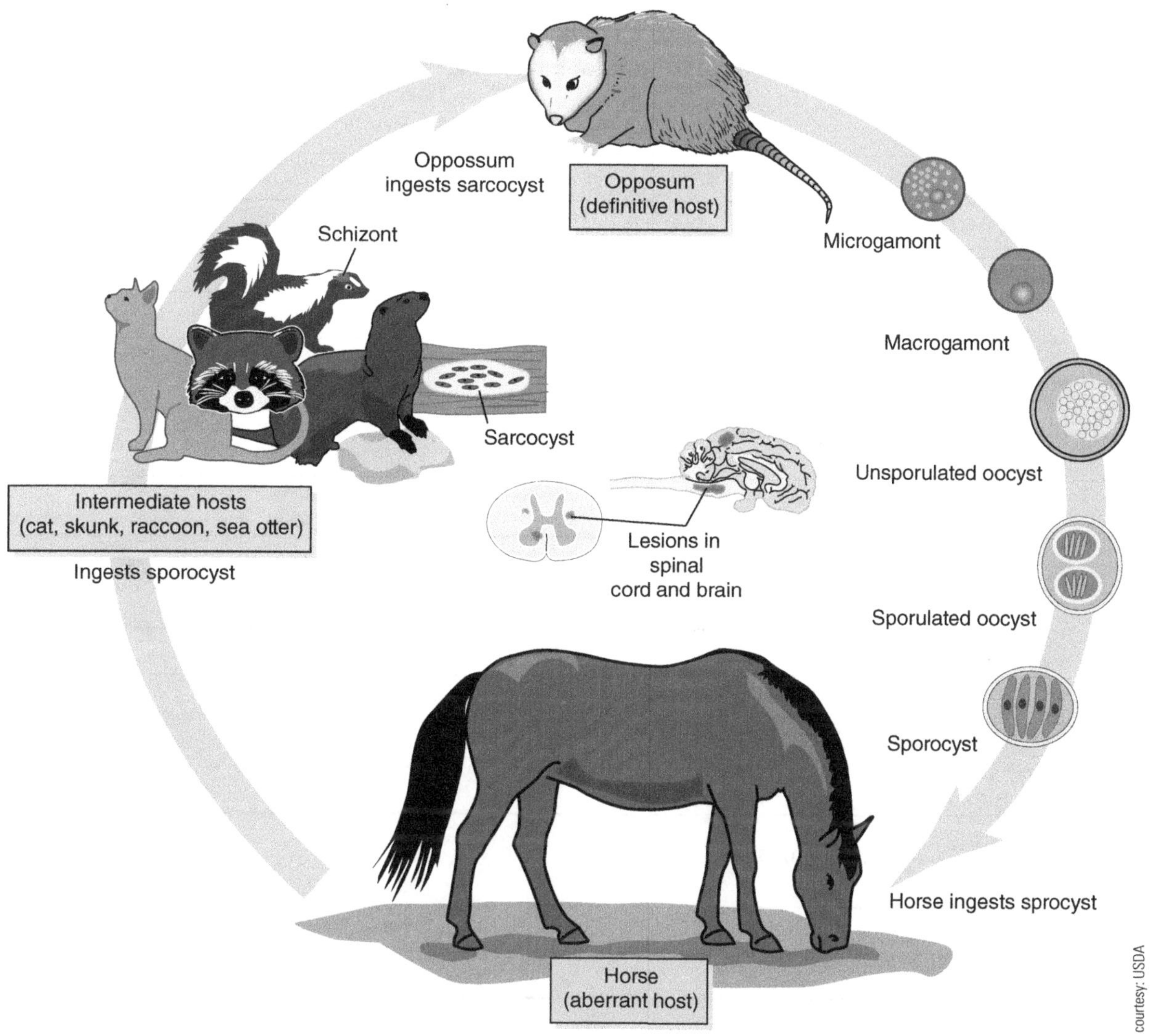

Figure 33-1 Equine parasite life cycle.

Directions: The instructor will divide the class into groups of two to three students each and then hand out index cards with a common parasite listed on each one. Examples are ascarids, small and large strongyles, and pinworms. Each group will need to answer the worksheet 1 questions based on the index cards. Next, the group should complete the parasite control worksheet 2.

Name____ Date__________________

Parasite Worksheet 1

The instructor will provide Playdoh. Present the findings to the class.

1. What is your parasite?

2. Describe your parasite. How big is it? What does it look like as an adult? Using Playdoh, make a model of your parasite to scale.

3. Where does your parasite live as a larva? As an adult?

4. What organs and other internal structures does your parasite damage during its life cycle? How?

5. List or draw the life cycle of your parasite.

6. List five things you can do to help prevent your pony from getting parasites.

Name_______________________________ Date_______________

7. Why is it important to deworm a new horse in the barn? What method would you use to deworm a new horse?

____

____

____

____

____

Parasite Worksheet 2

Complete the first chart with examples of three different types of parasite control. In the second chart, schedule the use of each parasite control for five different groups of horses.

Class	Brand	Active Ingredient
Avermectins		
Benzimidazoles		
Tetrahydropyrimidines		

Name____ Date____________________

Situation	Jan	Feb	Mar	Apr	May	June	Jul	Aug	Sep	Oct	Nov	Dec
A 10 horses/1000 acres												
B 20 horses/small pens												
C 2 horses/4 acres												
D 12 young horses/paddocks												
E 15 seniors												

A–Avermectin **B**–Benzimidazole **T**–Tetrahydropyrimidine

Name________________________________ Date______________

Chapter 33 Matching Activity

Directions: Match the term to the corresponding definition or description.

Term

____ 1. anhidrosis
____ 2. anthrax
____ 3. azoturia
____ 4. colic
____ 5. Cushing's syndrome
____ 6. distemper
____ 7. founder
____ 8. bot
____ 9. strongyle
____ 10. ascarid
____ 11. euthanasia
____ 12. encephalomyelitis
____ 13. equine infectious anemia
____ 14. heaves
____ 15. tetanus
____ 16. pinworms
____ 17. tapeworm
____ 18. vector

Definition

a. nutritional disorder that develops when a horse is put to work following a period of idleness
b. type of parasitic roundworm that infests the intestines of animals
c. disease caused by a small benign tumor in the pituitary gland
d. symptoms include high fever, loss of appetite, and depression; there is also pus-like discharge from the nose
e. flies that produce larvae that are parasites of horses
f. condition in which horses do not sweat normally
g. type of parasitic, blood-sucking worm that attacks the organs and tissues of animals
h. not a specific disease, but rather a disease complex encompassing a wide range of conditions that affect the horse's digestive tract
i. symptoms include swelling of the sensitive laminae on one or more feet, lameness, fever, and sweating
j. serious bacterial disease that is zoonotic
k. the practice of putting a sick or injured animal to death
l. internal parasites that may cause severe itching
m. nutritional disorder that affects the respiratory system
n. transmits from one organism to another
o. caused by a virus; it is carried from horse to horse by bloodsucking insects
p. disease that affects the brain; may be caused by any one of several different viruses
q. can cause unthriftiness and anemia
r. caused by bacteria; often enters the animal's body through a puncture wound but may enter through other types of wounds

Name______________________________ Date______________

Chapter 33 Lab Questions

1. What are the results if a horse is untreated for parasites and disorders?

2. What is the most effective strategy to control disorders and parasites?

3. List the most common equine disorders.

4. What are some general symptoms of internal parasites in horses?

Feel good studio/Shutterstock.com

Chapter 34

Training and Horsemanship

Introduction

Training a young horse is a delicate, slow, and detailed process. Each skill that riders usually take for granted must be trained from the beginning. Experienced trainers receive a premium fee for their services because working and training horses is a difficult job and requires knowledge of equine psychology, patience, and repetition. A horse that cannot be trained loses value at a fast rate. Understanding the reasons for horse behavior is helpful when training and riding horses. Horses must move their heads to bring objects into focus, so they naturally shy away at sudden movements. Horses have good memories, making it possible to train them to respond to commands, but negative experiences also remain in their memory and are hard to correct. Sensitive areas on the body of the horse are used to help control the animal, so a relationship of trust must be established between the trainer and horse.

Groundwork is the first and most essential part of a horse's training. Horses need to develop basic habits such as lifting their feet for farrier work and being led from place to place. Some basic groundwork that must be done when "breaking" or training a young horse includes haltering, leading, working with feet, saddling, longeing, long reining, neck-reining, and pole work. Once the groundwork is established, training with the saddle may begin. The use of a hackamore and a bridle with a snaffle bit are strategies to maintain control of the untrained horse without causing pain or discomfort. When mounting a horse for the first time, special precaution must be taken to ensure safety to the horse and rider.

As the use of horses for work and transportation has decreased in the United States, horse equitation has gained in popularity. Equitation classes are judged on the performance of the rider.

The rider must have skill in riding and controlling the horse. Position in the saddle, use of the hands, proper tack and dress, and the performance of the various gaits are the main points judged in equitation classes.

In this exercise, complete the young horse training activity and explore various equitation events available for horse and rider to compete against other horse and rider pairs.

© Appaloosa Horse Club

Name____ Date__________________

Exercise 34-1 Training a Young Horse Activity

Young horses should be handled every day to repeat basic groundwork tasks. At the beginning, the time spent with young horses should not exceed 5 to 10 minutes for each training session.

Directions: For each of the training situations, describe three important training procedures required for proper training of the desired task. For each step, include a picture or drawing of the horse and trainer performing the task. In the space provided, draw each step or find a picture that can be placed in the space provided. Training procedures should include drawings or pictures.

Haltering

1. ____________________
2. ____________________
3. ____________________

Leading

1. ____________________
2. ____________________
3. ____________________

Working with the feet

1. ____________________
2. ____________________
3. ____________________

Longeing

1. ____________________
2. ____________________
3. ____________________

Saddling

1. ____________________
2. ____________________
3. ____________________

Name____ Date__________________

Using a hackamore and bridle

1. ____
2. ____
3. ____

Driving

1. ____
2. ____
3. ____

Name____ Date____________________

Exercise 34-2 Equine Performance Events

Equine performance events have grown in popularity as many horse owners compete as a hobby or pastime. These events serve as spectator events and a form of social interaction for riders and families that participate.

Directions: For each of the following equine events, (1) describe the objectives and rules of the event; (2) describe the appropriate equipment for the event, including tack and proper clothing for the rider; and (3) provide two examples of horse breeds that are commonly used for this event.

Horse evaluation

1. ____
2. ____
3. ____

English equitation

1. ____
2. ____
3. ____

Gymkhana

1. ____
2. ____
3. ____

Rodeos

1. ____
2. ____
3. ____

Name______________________________ Date________________

Chapter 34 Matching Activity

Directions: Match the term to the corresponding definition or description.

Term	Definition
____ 1. longeing	**a.** when a rider rises and sits in the saddle
____ 2. horsemanship	**b.** controlling the response of the horse by the weight of the rein against the neck
____ 3. neck-reining	**c.** training the horse at the end of a 25- to 30-foot line
____ 4. posting	**d.** term used for games on horseback
____ 5. gymkhana	**e.** art of riding a horse
____ 6. imprinting	**f.** handling a foal each day while it is young to bond with the handler
____ 7. sacking out	**g.** two types of horsemanship are Western and English
____ 8. equitation	**h.** a coiled rope
____ 9. shown at halter	**i.** means horses are led, not ridden, and are judged for conformation, soundness, and action
____ 10. riata	**j.** rubbing a sack or saddle pad over the head, neck, back, rump, and legs

Name______________________________ Date______________

Chapter 34 Lab Questions

1. What are the basic, natural behaviors of a horse that a trainer must understand?

2. What are some groundwork practices? Why are these important?

3. Describe the process of mounting a horse for the first time.

4. What is equitation? Why is it popular in the United States?

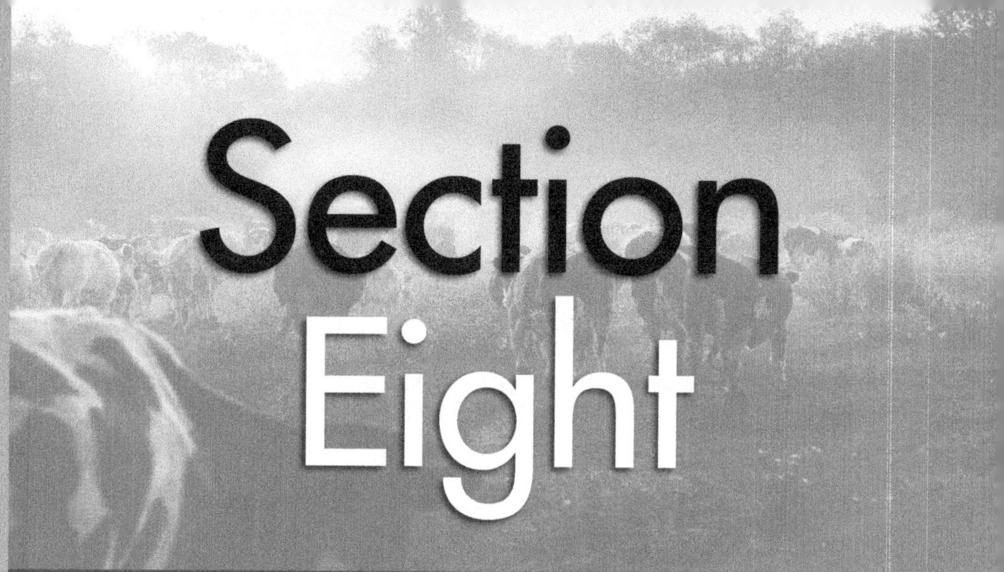

Section Eight

Poultry

Feel good studio/Shutterstock.com

Feel good studio/Shutterstock.com

Selection of Poultry

Introduction

Besides niche markets and small-scale operations, most of the poultry industry consists of large commercial production flocks. The two major types of poultry enterprises are egg production and raising pullets for replacement or meat production. The majority of the poultry industry consists of chicken production—either broilers for meat or layers for egg production. Turkeys, ducks, and geese are raised for meat in the United States but make up only a small part of the industry. A large percentage of the industry is vertically integrated in order for the production companies to control the product quality from production to the consumer.

The production of poultry has fluctuated over the past four or five decades but has seen a steady increase in recent years. Although the production of eggs has increased, the consumption per capita has decreased. Iowa, Indiana, Ohio, Pennsylvania, and Texas are the leading states for laying hens and egg production. Georgia, Arkansas, Alabama, North Carolina, and Mississippi are the leading states in the production of chicken broilers. There has been a major increase in broiler production and per capita consumption of poultry since 1970.

Most chickens used for egg production in the United States are Leghorn strain crosses. Crossbred chickens are generally used for broiler production because of accelerated growth rates. In either section of the industry, chickens are bred for the type of production desired. Niche market poultry producers utilize purebred chickens more often to differentiate their product. The breed most used for production turkeys is Broad Breasted Large Whites. The White Pekin duck is most commonly used for meat production, and the most popular breeds of geese are the Toulouse and the Embden, used primarily for meat production.

Poultry enterprises must purchase replacement birds on a regular basis; they must also regularly cull older birds that decrease in production. Most replacement birds are purchased from hatchery operations as day-old chicks, started birds, or ready-to-lay chickens. Culling is based on appearance and condition of the body. Pigmentation and molt are indicators of birds losing production quality for egg-laying production.

This chapter explores the poultry industry as well as the anatomy of the chicken.

Name____ Date____________________

Exercise 35-1 Growth of the Poultry Industry

The poultry industry has fluctuated over the past five decades due to many different factors. Over recent years, the trend in poultry has been an increased number of production chickens. In this exercise, assume the role of a poultry industry consulting firm. A client is looking to start up a poultry operation and needs assistance understanding the historical trends of the market, identifying the current market conditions, and choosing the best location and type of enterprise to enter the industry. Groups of students will analyze the poultry market and suggest the best strategy for this client to enter the industry.

Directions: The instructor will divide the class into consulting teams. As a team, create a market analysis for the perspective poultry producer. The market analysis should include the following information:

- Describe the different poultry enterprises that are commonly used.
- Explain vertical integration.
- Describe the trends in poultry production and consumer consumption.
- Explain the recent trends for broilers and for egg production.
- Give a short description of the turkey industry trends for comparison.
- Suggest a location for the operation based on the records of production in various states.
- Describe the advantages and disadvantages to raising poultry.
- Outline the various breeds that can be used and list their characteristics.
- Suggest sources where the client may obtain chickens for production.
- Summarize a suggested strategy for this client to enter the poultry industry.

The resulting market analysis should be written as a proposal to be handed in to the instructor or as a presented proposal using a computer-based presentation program.

Name______________________________ Date______________

Exercise 35-2 Chicken Anatomy Activity

Locating and learning the poultry parts on a live bird are important skills in the industry. Before real birds are used, this exercise will use a diagram to identify the parts of a chicken in Part 1, and Part 2 includes a chicken wing dissection to better understand the internal anatomy.

Part 1

Directions: Using the following word bank, identify the external parts of a chicken.

Word Bank:

Beak	Breast	Comb	Ear Lobe
Hock	Eye	Shank	Shoulder
Toes	Vent	Waddle	

Name______________________________ Date______________

Part 2: Chicken Wing Dissection

Through this dissection activity, make sure to observe the muscle, bones, and blood vessels that make up a chicken's wing. The instructor will provide the needed materials for the dissection.

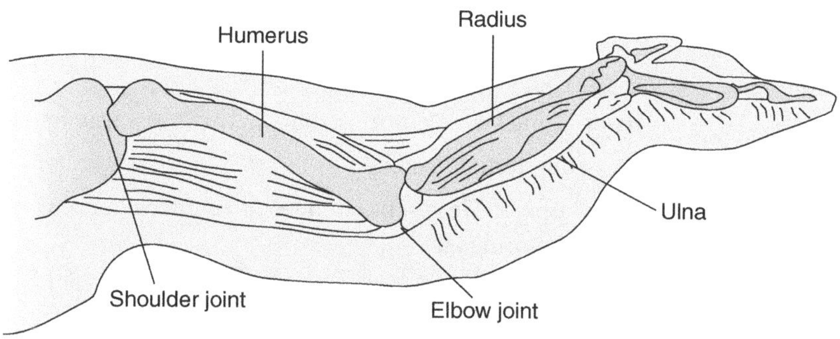

Dissection Instructions

1. Rinse the chicken wing under cool water and then dry it off.
2. Using scissors, cut the skin down the middle of the wing from where it was removed from the body toward the tip of the wing. (The scissors should cut the skin but not the muscle.)
3. Next, cut a line across the first line, approximately in the middle of the first cut.
4. Peel back the layers of skin to reveal the muscle, tendons, and bones.
5. Observe all the internal parts of the chicken wing and complete the following table with the information that is gathered.

Tissue	Description (color, texture, etc.)	Tissues it is attached to
Skin		
Fat		
Muscle		
Tendon		
Ligament		
Cartilage		

Conclusion

Based on the observations, explain the roles of muscles, bones, tendons, and joints in the movement of the lower chicken wing.

Name____ Date____________________

Chapter 35 Matching Activity

Directions: Match the term to the corresponding definition or description.

Term

____ 1. vertical integration
____ 2. culling
____ 3. variety
____ 4. cross-mating
____ 5. hybrid
____ 6. comb
____ 7. pinfeather
____ 8. drake
____ 9. straight run chick
____ 10. pullet
____ 11. broiler
____ 12. class
____ 13. breed crossing
____ 14. inbreeding
____ 15. bantam
____ 16. hatchability
____ 17. cockerel
____ 18. sexed chick
____ 19. molting
____ 20. poult
____ 21. trap nest
____ 22. candling

Definition

a. breeds that are a direct result of breed crossing; crosses are also popular for egg production
b. crossing two or more strains within the same breed
c. young female chicken
d. groups of birds separated by differences in color of plumage, type of comb, and size
e. male duck
f. process of removing undesirable chickens from the flock
g. about one-half pullets and one-half cockerels
h. where two or more steps of production, marketing, and processing are linked together
i. feather that is not fully developed
j. examples include single, rose, pea, cushion, buttercup, strawberry, V-shaped, and walnut
k. raising chickens for meat
l. nest equipped with a door that automatically closes when the bird enters the nest
m. geographic origin of the poultry
n. young male chicken
o. breeding of closely related individuals within a breed to get the desired traits
p. number of young
q. process of holding a light against the egg to observe the embryo
r. process of losing the feathers from the body and wings
s. pullets and cockerels
t. young turkey
u. crossing different breeds to get the desired traits
v. small chickens, usually weighing from 16 to 30 ounces as adults

Name______________________________________ Date__________________

Chapter 35 Lab Questions

1. What recent changes have occurred in the poultry industry? Why do you think they have occurred?

2. Explain the differences between a broiler and a laying hen.

3. What are the advantages and disadvantages of raising poultry?

Feel good studio/Shutterstock.com

Chapter 36

Poultry Feeding, Management, Housing, and Equipment

Introduction

Poultry require carbohydrates, protein, minerals, vitamins, and water to grow and remain healthy. The highest cost in poultry production is the expense of feed. Smaller flock owners generally find it easier to use a complete commercial feed, while large commercial flock owners usually use some system of mixing ingredients to make a complete feed. Two ingredients commonly added to chicken feed are grit and oyster shells. Grit is used to help the bird grind the feed in the gizzard, and oyster shell provides calcium for egg production.

Water is an essential part of a poultry operation. Controlling the amount and quality of water can directly affect the flock's profitability. Understanding the expected water consumption rates and the acceptable levels of compounds in the water is essential when raising poultry. Light and temperature have a significant effect on the growth of poultry. Light affects the sexual maturity and egg production rates of poultry. Proper temperatures are required for successful brooding, meat production, and egg production. Many poultry production buildings are equipped with tunnel ventilation fans to maintain an adequate temperature.

Poultry may be brooded on the floor or in cages. Chickens raised for meat production are usually raised in confinement, unless supplying a niche market such as for free-range chicken. Laying flocks may be handled using floor systems or in cages. Automation has been utilized in a majority of laying operations to reduce labor and human error. Eggs must be carefully handled to maintain quality.

Housing and equipment requirements differ greatly depending on the size, age, species, and breed of poultry, as well as the purpose for which they are raised. Turkeys, geese, and ducks generally require less investment in facilities and equipment. Automatic or semiautomatic feeding, watering, and cleaning equipment are frequently used in the commercial poultry industry. The standards demanded through vertical integration require most poultry producers to eliminate any areas of inefficiency in the operation.

This chapter provides two exercises that develop an understanding of poultry requirements.

Name____ Date__________________

Exercise 36-1 Water Consumption Lab

Water is a critical nutrient that receives little attention until a problem arises. Not only should producers try to provide water in adequate quantity, but they should also know what is in the water that the poultry will be consuming. In this three-part exercise, students will study the water consumption levels of birds at different sizes and stages of life, as well as analyze the quality of the water being provided to the poultry.

Part 1

Directions: Use information from the textbook to determine how much water should be consumed each day by different groups of chickens. For each of the following types of chicken and groups, determine the level of water consumption that is needed to keep the entire flock hydrated.

A flock of 500, 1-week-old broilers:

A flock of 100, 4-week-old white leghorn hens:

A flock of 400, 1-week-old brown hens:

A flock of 5,000, 8-week-old broilers:

A flock of 2,000, 16-week-old white leghorn hens:

A flock of 1,500, 20-week-old brown hens:

Name____________________________________ Date________________

A flock of 750, 4-week-old broilers:

A flock of 2,500, 1-week-old brown hens:

Part 2

The instructor will provide several samples of live birds. There will be an assortment of birds of different ages, sizes, and breeds. A total of six different birds should be used. Separate each bird into a distinct area with its own feed and water. Monitor each bird's feed and water intake. Use a watering container with measurement markings for easy data recording.

Directions: Monitor and record the data for each bird and answer the following questions.

1. Determine the bird that requires the most water intake.
2. Why does the bird with the highest intake consume so much water?
3. How much feed did the bird consume?
4. What is the bird's feed-to-water ratio?
5. What are some environmental factors that could have affected the water consumption rates?

Name______________________________ Date______________

Part 3

The quality of the water provided to poultry is just as important as providing the adequate amount. In this exercise, use a standard water quality test provided by the instructor to determine the levels of various substances in the water.

Directions: Table 36-1 is a guideline for acceptable levels of compounds found in drinking water. Collect several samples of water from the waterers located on a sample poultry operation. Compare the levels of compounds in the drinking water with the levels given in the guidelines. Create an analysis report and determine if the water in the samples is acceptable for poultry to consume. Present your findings to the instructor and the class.

Table 36-1

Characteristic or Mineral	Maximum Acceptable Levels
pH	6.0–8.0
Hardness	110 ppm
Naturally Occurring Compounds	
Calcium	500 ppm
Chloride	250 ppm
Copper	0.6 ppm
Iron	0.03 ppm
Magnesium	125 ppm
Manganese	0.05 ppm
Nitrate	25 ppm
Phosphorus	0.1 ppm
Potassium	500 ppm
Sodium	50 ppm

Name_______________________________________ Date___________________

Exercise 36-2 Management Lab: Broilers vs. Layers

Managing chicken operations for meat production and egg production requires two very different strategies. The requirements for feed, housing, equipment, animal health, labor, growth, and product processing in a laying operation are completely different from those for a meat production operation.

Directions: In the space provided, highlight the major differences in management requirements for an egg production operation and a meat production operation. Include specific requirements recommended in the textbook. Also include any similarities in requirements.

Egg production system requirements:

Meat production system requirements:

List any similarities between egg and meat production systems:

Name____________________________ Date______________

Chapter 36 Matching Activity

Directions: Match the term to the corresponding definition or description.

Term

____ 1. ad libitum
____ 2. best management practices
____ 3. condemnations
____ 4. brooder
____ 5. battery cage
____ 6. grit
____ 7. negative pressure system
____ 8. positive pressure system
____ 9. phase feeding
____ 10. feeding efficiency
____ 11. poult
____ 12. hover guard
____ 13. beak trimming
____ 14. foot-candle
____ 15. capon
____ 16. broodiness
____ 17. tunnel ventilation
____ 18. dry cleaning
____ 19. snood
____ 20. spraddled legs

Definition

a. the number of pounds (kilograms) of feed required to produce a dozen eggs
b. a male chicken that has been surgically castrated
c. a young turkey
d. when chickens are given all they will eat
e. the cutting off of one-third to one-half of the upper beak and one-fourth of the lower beak
f. a unit of luminance on a surface that is 1 foot from a point source of a candle or light
g. used for the first week of brooding to prevent the chicks from wandering away from the heat and becoming chilled
h. a system of making specific feeds to be used to meet the changing nutritional requirements of chickens
i. when a hen stops laying eggs and wants to sit on a nest of eggs to hatch them
j. small particles of granite; comes in small, medium, and large sizes for use with chicks, growing chickens, and adult chickens, respectively
k. necessary for managing feeding, housing, equipment, and general production needed for a profitable enterprise in modern poultry production
l. birds determined to be unfit for human consumption
m. infrared heat lamps
n. if fans are used to exhaust the air from the building
o. uses air inlets at one end of the building and exhaust fans at the other end to move air through the building at a minimum velocity of about 350 feet per minute
p. systems using layers of two or more cages stacked upon each other
q. if fans are used to bring fresh air into the building
r. buffing the eggs with fine sandpaper, emery cloth, or steel wool
s. a condition in which the leg is dislocated from the socket where it is attached to the body
t. a tubular fleshy appendage on the top of the head and helps to prevent head injuries from pecking or fighting

Name______________________________ Date______________

Chapter 36 Lab Questions

1. What are the major requirements for raising poultry?

2. How does light affect poultry production?

3. What are three types of confinement structures for poultry? Explain each one.

Feel good studio/Shutterstock.com

Diseases and Parasites of Poultry

Introduction

To maintain a good health program for poultry, a producer must focus on various methods of disease prevention. The last resort should be treatment, which can be costly and is never guaranteed to help the birds recover. The best way to control disease and the infestation of parasites is to create a prevention plan. Prevention involves sanitation, good management, vaccination, and control to stop and treat disease outbreaks. When poultry animals become sick, they rarely recover. Treatment is only effective when the sickness is caught and treated early. Without a proper treatment plan, the profits of a poultry operation decrease significantly. The involvement of a veterinarian is recommended when developing a treatment plan or diagnosing an observed disease or parasite.

To properly sanitize during disease prevention and control, all poultry buildings and equipment should be disinfected and scrubbed free of any lingering litter from previous groups of birds. Allowing a dormant period for the housing and equipment to dry out between poultry groups is also recommended. Any birds that become infected should be removed from the flock and isolated where they cannot come in contact with healthy birds. Any dead birds should be disposed of, and manure from infected birds should not encounter healthy birds in the flock.

Other methods to prevent the outbreak of disease include vaccinating young birds and checking daily for signs of disease or parasite. A drop in production or an increase in the death rate may indicate the presence of a disease. Even veterinarians cannot always accurately diagnose an illness in poultry, so they may rely on laboratory tests to assist in a diagnosis.

There are many diseases that affect poultry. A large number have similar symptoms and effects, and there are several medicines that treat a wide variety of sickness. Some diseases are specific to the species of poultry, while many sicknesses are common to other poultry species. All poultry are affected by mites and lice if not properly treated. Maintaining a sanitary environment and the proper use of insecticides can help control the spread of these parasites.

Name____ Date____________________

Exercise 37-1 Disease and Parasite Control Program Activity

Understanding medication labels and properly administering them to the animals is an important skill. In this exercise, students will calculate the first day that any chickens can safely be marketed for food, calculate the correct amount of medication to mix with water when making stock solution, and calculate the amount of stock solution to mix with the chickens' drinking water.

Directions: Read the label instructions provided in Figure 37-1. Use the label and the following situation text to answer all the questions about administering the medication.

Carramycin-152

(oxytetracycline HCl as soluble water)

For control and treatment of specific disease in poultry, cattle, swine and sheep.

CAREFULLY READ ALL DIRECTIONS BEFORE USING THIS PRODUCT.

Soluble Powder for Use in Drinking Water Only.

Active Ingredients: Carramycin-152 is a broad spectrum **antibiotic.** This 4.78 oz packet contains 102.4 grams oxytetracycline HCl (after mixing with clean, fresh water-512 gallons containing 200 mg oxytetracyline HCl per gallon. 256 gallons containing 400 mg oxytetracycline HCl per gallon: 128 gallons containing 800 mg oxytetracycline HCl per gallon).

Indications: For control of **poultry** diseases caused by organisms susceptible to oxytetracycline.

Recommended Dosage

Add the following amount to two (2) gallons of fresh, clean water to make a stock solution. Mix one (1) ounce solution per one (1) gallon drinking water.

		Dosage	***Packs/2 Gallons Stock solution***
Chickens	Infectious synovitis	200 mg/gal	1/2
	Chronic respiratory disease	800 mg/gal	2
	Fowl Cholera	800 mg/gal	2
Turkeys	Infectious synovitis	400 mg/gal	1
	Hexamitiasis	200 mg/gal	1/2

Cautions: 1. Carramycin-152 is for use in flock drinking water only. 2. Medicate continuously at the first clinical signs of disease and continue for 7 to 14 days. If improvement is not noticed within 24 to 48 hours, consult a veterinarian or diagnostic laboratory to determine diagnosis and advice on dosage. 3. Use as sole source of oxytetracycline. Do not use for more than 14 consecutive day in chickens and turkeys or five (5) consecutive days in cattle, sheep or swine. 4. Carramycin-152 is to be stored below 77°F. 5. The concentration of drug required in the medicated water must be adequate to compensate for variations in age of the animal, feed consumption rate, and the environmental temperature and humidity-each of which affects water consumption.

Warning: Do not administer to chickens, turkeys, wine, cattle, or sheep within five (5) days of slaughter. Do not administer to chickens or turkeys producing eggs for human consumption.

How Supplied: Carramycin-152 soluble powder is available in packets of 4.78 oz.

Livestock Drug – *not for human use*

KEEP OUT OF REACH OF CHILDREN

Distributed by

Livestock Health, Inc.

Source: Poultry Learning Laboratory Kit, © 1998, Curriculum Materials Service, Ohio State University

Figure 37-1 Example of a pharmaceutical label.

Name______________________________ Date______________

Situation

As the owner of a poultry meat production operation, you care for the health of 3,200 red broilers. The flock health manager reports that most of the chickens are coughing and having difficulty breathing. After calling the veterinarian in to look at the flock, they are diagnosed with a chronic respiratory infection. The good news is that this is a treatable infection. The veterinarian sells you a water-soluble medication that can be administered to the flock through their drinking water. The medicine is Carramycin-152. You instruct the manager to immediately begin mixing the powder into the water and keep track of the time period the medicine is administered; the start date is May 28. The veterinarian recommends treating the birds with the soluble powder mixed in the drinking water for 10 days. The directions on the medication label direct you to mix 800 mg of Carramycin for every gallon of water. The flock of broilers drinks a total of 256 gallons of water each day; for each gallon of water, mix one ounce of the soluble powder.

Medication Questions

1. How many packs of this medicine should be mixed with two gallons of water?

____

____

____

2. How much soluble powder is needed for 10 gallons of water?

____

____

____

3. How many packs of Carramycin will be needed to treat the broilers for 10 days?

____

____

____

4. How many ounces of the soluble powder will be needed each day?

____

____

____

Name______________________________ Date______________

5. How many milligrams of the medicine will each chicken in the flock receive?

6. What is the first day the broilers could be slaughtered when all of the residue from the Carramycin is gone? Give a date.

Name____________________________________ Date________________

Chapter 37 Matching Activity

Directions: Match the term to the corresponding definition or description.

Term	Definition
____ 1. intranasal vaccination	**a.** the process of injecting the vaccine into the skin on the underside of the wing web at the elbow
____ 2. intraocular vaccination	**b.** a transparent lubricating fluid in the joints of poultry
____ 3. wing web vaccination	**c.** placement of a vaccine directly into the nose opening
____ 4. synovia	**d.** a vaccine made with weakened bacteria
____ 5. bacterin	**e.** placement of the vaccine directly into the eye
____ 6. virulence	**f.** the severity of the disease

Name______________________________ Date______________

Chapter 37 Lab Questions

1. What are some of the major causes of disease and parasites in poultry?

2. What is the most effective strategy to control infection from disease and parasites?

3. List the most common diseases in poultry.

4. What are the most common parasites to affect poultry?

5. Describe the life cycle of parasites and how they enter the body of the animal.

Marketing Poultry and Eggs

Introduction

The poultry market has fluctuated somewhat over the past 40 years, but in general, consumption of broilers and turkeys has shown an upward trend. Egg consumption has been decreasing over this period, but the trend has been rebounding in recent years. Long-term trends in prices show variations resulting from supply and demand and the influence of vertical integration on the market. Vertical integration is a system in which the company that markets the poultry also owns the poultry through the entire process of raising from chick to market animal.

Most of the large chicken brands that are recognizable across the nation in supermarkets are using vertical integration. In the United States, 99 percent of the broilers are produced under some type of vertical integration contract. At the large volumes that these integrated contracts are marketing, the poultry must be carefully handled during every step to avoid injuries and bruises. Most of the turkeys in the country are marketed through integrated contracts or cooperatives, while ducks and geese are more commonly processed on the farm and directly marketed to consumers.

Each species of poultry has carcass classes and grades to assess the quality of meat. Classes are based on age, tenderness of meat, smoothness of skin, and hardness of breastbone cartilage. These "ready-to-cook" carcasses are classed and then graded. The three grades are known as A, B, and C. These grades and all standards for grading poultry are found in the U.S. Department of Agriculture (USDA) Poultry Grading Manual.

A large section of the poultry egg industry is marketed though vertical integration as well. The rest of the egg market includes sales to local buyers, produce dealers, cooperatives, or direct sales to consumers. Eggs are classed and graded to measure their quality. Eggs must be handled carefully to maintain high quality because defects due to handling can decrease the value of an egg or even render it unsellable. The USDA also sets the standards for weight classes and grades of eggs. The egg classes are based on the weight in ounces per dozen eggs. The grades are based on shell quality, size of the air cell, firmness of the egg white, and the presence of defects in the yolk. The ability to grade and class eggs is necessary in the poultry industry.

© Daleen Loest/Shutterstock.com

Candling eggs to determine their grade.

Name________________________________ Date______________

Exercise 38-1 Egg Grading Lab

Directions: Answer all of the following questions about quality grading eggs. Once the questions are complete, the instructor will provide a sample of various qualities of eggs. Use Table 38-1 in the textbook for quality grade specifications. Grade each sample egg and explain why it received that grade.

Grading cartons of eggs

1. Name three factors used in grading a carton of eggs.

____

____

____

____

2. What is the difference between a checked and a cracked egg?

____

____

____

____

3. What is a body check?

____

____

____

____

4. Grade example cartons of eggs provided by the instructor and record the grades and reasons.

____

____

____

____

Name______________________________ Date__________________

Interior egg evaluation

1. What process is used to evaluate the interior of an egg?

2. What popular comparison is used when determining the size of an egg's air sac?

3. What common defects are evaluated when grading the interior of an egg?

4. Describe the characteristics of the yolk from a freshly laid egg.

5. Candle several sample eggs and determine the interior quality of the egg.

6. Determine the grade of egg based on its break-out quality.

Name______________________________ Date______________

Shell evaluation

1. What are the four grades of eggshells?

2. Describe the six factors used to determine eggshell quality.

3. What will automatically cause an egg to receive a grade of B?

4. What two defects cause a grade of dirty?

5. Grade the shells of all sample eggs. Explain the reasons for each grade.

Name________________________________ Date__________________

Chapter 38 Matching Activity

Directions: Match the term to the corresponding definition or description.

Term	Definition
____ 1. viscera	**a.** eggs that are examined by using a high-intensity light
____ 2. green geese	**b.** internal poultry organs
____ 3. candling	**c.** geese that are full fed for fast growth
____ 4. marketing	**d.** a widely held but false belief
____ 5. myth	**e.** involves processing, promoting, selling, and distribution of products

Name____ Date__________________

Chapter 38 Lab Questions

1. Describe the long-term trends in the poultry market.

2. What system is used for the majority of poultry marketing?

3. What qualities are "ready-to-cook" poultry carcasses graded on?

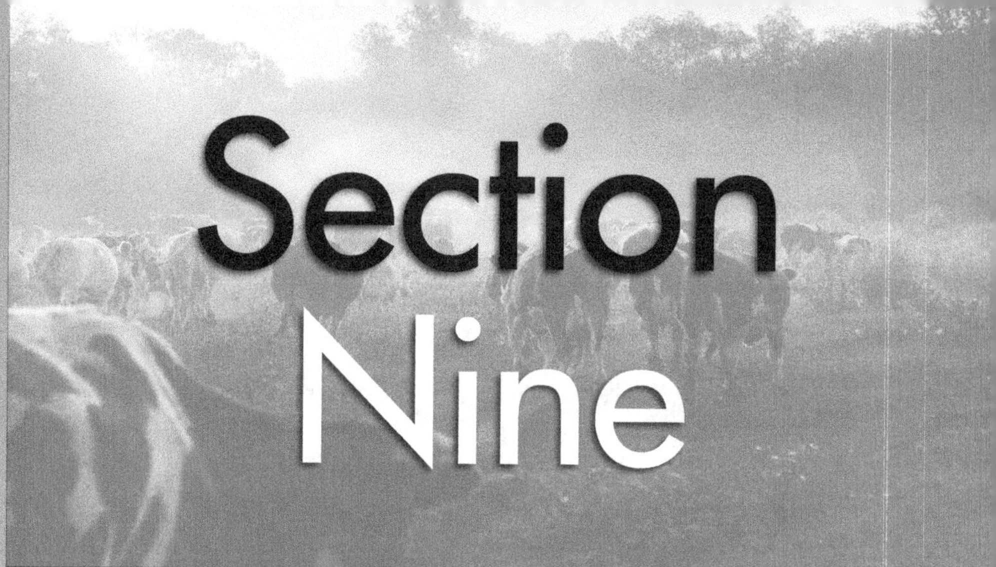

Section Nine

Dairy Cattle

Feel good studio/Shutterstock.com

Feel good studio/Shutterstock.com

Chapter 39

Breeds of Dairy Cattle

Introduction

Many of the dairy breeds used in modern agriculture originated in Europe. The cattle were selected for the most desirable traits for the area. The most desirable animals were kept as breeding stock. The development of dairy breeds has continued into the current dairy industry. Herds that have been managed for generations have achieved a high level of genetics that receive premium value. Dairy cows that produce large amounts of milk or high levels of milk components are desired. Cows are constantly producing milk during lactation, so they are milked two to three times per day, seven days a week, continuously.

The dairy industry is very difficult—and labor intensive. It also requires a large capital investment to begin a dairy operation; many commercial operations are generational family farms or large farms owned by investors. Extensive training is preferred for those managing dairy cattle. Most dairy farms hire a herd health manager; for this position, a college education is desirable.

The number of dairy farms in the United States has been declining in recent years, but the numbers of dairy cattle have recently been increasing slightly. The trend has been for small herds of dairy cattle to leave the industry and for the large commercial dairy farms to expand and grow. Due to this trend, the average size of the dairy farm has increased. Improvements in herd health, genetics, and milking equipment technology have caused the milk produced per cow to increase as well.

The seven major breeds of dairy cattle in the United States are Ayrshire, Brown Swiss, Guernsey, Holstein-Friesian, Jersey, Milking Shorthorn, and Red and White. There are breed associations for the purpose of registering purebred dairy cattle. Registry ensures the animal's pedigree and increases its genetic value. Dairy cattle have been developed over the generations by selecting for quantity and persistence of milk production. Each breed is selected based on the characteristics commonly found in the breed, but when interested in milk production, a farmer should milk any breed of cattle that is profitable. The following exercises are designed to teach students basic breed identification skills and an understanding of the dairy industry.

Name____ Date____________________

Exercise 39-1 Trends in Dairy Production Activity

Create a timeline that highlights the various trends in the dairy industry using the graphs in Chapter 39 of the textbook. The timeline should track the changes in the dairy industry that have taken place from 1975 to 2010. The timeline should describe the conditions of the dairy industry during that year. The categories to include on the timeline are (1) the milk cow numbers, (2) total milk production, (3) milk production per cow, (4) production of milk fat per cow, (5) the consumption of fluid milk and cream, (6) consumption of milk and nonfat dry milk, (7) consumption of cheese and ice cream, and (8) the consumption of dry whole milk.

Directions: Create a timeline of the changes in the dairy industry since 1975. Include each category of dairy industry measurements that apply. Include the known reasons for the changes that occur.

Name____ Date____________________

Timeline of Dairy Industry

Date	Changes in the Industry (industry data)

Name____ Date____________________

Exercise 39-2 Breed Comparison Lab

Part 1

Directions: Each of the following pictures shows a different dairy cattle breed. For each picture, identify the breed and write a brief history and description of the breed. Include any unique characteristics of the breed.

1. 

© Ayrshire Breeder's Association

a. Breed: ____
b. History: ____
c. Description: ____

2. 

© Agri Graphics Ltd

a. Breed: ____
b. History: ____
c. Description: ____

3. 

© Brown Swiss Cattle Breeder's Association of America

a. Breed: ____
b. History: ____
c. Description: ____

Name________________________________ Date______________

4. 

Holstein Association USA, Inc.

a. Breed: ______________________________

b. History: ______________________________

c. Description: ______________________________

5. 

Photo by Cybil Fisher. Reprinted with permission of the American Milking Shorthorn Society

a. Breed: ______________________________

b. History: ______________________________

c. Description: ______________________________

Part 2

Directions: Compare breeds based on their use in production. Determine which breeds would be suitable to (1) large commercial operations that milk three times per day, (2) niche farms that produce alternative dairy products, (3) registered purebred herds, or (4) small-scale farms that sell dairy products directly to the consumer.

Brown Swiss

Holstein-Friesian

Ayrshire

Guernsey

Red and White

Jersey

Name__________________________________ Date__________________

Chapter 39 Matching Activity

Directions: Match the term to the corresponding definition or description.

Term

____ 1. Grade A milk
____ 2. Grade B milk
____ 3. registered
____ 4. milking parlor
____ 5. breed standard
____ 6. breeding true
____ 7. composite breed
____ 8. foundation stock
____ 9. heterosis
____ 10. homogeneous
____ 11. purebred
____ 12. seedstock
____ 13. trait
____ 14. herdbook
____ 15. grade animals
____ 16. DHIA
____ 17. dystocia
____ 18. roan

Definition

a. eanimals that meet the requirements of the breed association and are recorded in the association's herdbook
b. milk that is produced on farms that have been certified to meet certain minimum standards
c. milk that is produced under conditions that are less controlled than those for producing Grade A milk
d. a very close mixture of two different hair colors
e. the tendency of crossbred offspring to possess traits superior to the parents, also known as hybrid vigor
f. a barn where dairy cattle are milked daily
g. calving difficulty
h. distinguishing characteristics such as color and size within a specific breed
i. a written description of desirable and undesirable breed characteristics
j. registered breeding cattle of high merit to supply genetics to commercial producers
k. a breed made up of two or more other breeds and maintained as purebreds
l. an animal whose parents are of the same breed
m. animals directly ancestral to a breed
n. ability to pass characteristics to offspring in a highly consistent manner
o. sharing the same or very similar traits
p. the official records of individual and pedigrees of a breed
q. Dairy Herd Information Association
r. the result of mating a registered bull with a nonregistered cow or one of mixed breeding.

Name____ Date____________________

Chapter 39 Lab Questions

1. What qualities of milk are desired other than quantity?

____

____

____

____

2. What is the difference between Grade A and Grade B milk?

____

____

____

____

3. What are the desirable qualities of a milking shorthorn?

____

____

____

____

4. What would be considered a small dairy farm in the United States?

____

____

____

____

Feel good studio/Shutterstock.com

Chapter 40

Selection and Judging Dairy Cattle

Introduction

Judging dairy cattle gives a general idea of the qualities that a cow has. The more judging events an animal wins or performs well in, the more valuable its genetics will be. A good judge learns the parts of the animal and develops a logical system for evaluating the animal using the Dairy Cow Unified Scorecard.

Physical appearance, health, milk production records, and pedigree are important factors to consider when selecting dairy animals for breeding and production. The mammary system is one of the most important parts of the dairy cow and carries the most weight when scoring and evaluating the cow.

When dairy cows are being evaluated, a judge should have a routine for visual inspection to ensure an accurate comparison between cattle. A dairy cow should be observed from the side, rear, and front. A judge should have experience and credibility in the cattle industry in order to build a reputation among dairy producers. The judge should give a presentation of oral reasons and use the proper terms to describe cattle conformation.

Producers have many strategies to best prepare a cow for a dairy cattle show. Feeding and water strategies prior to the judging event can affect the way the cow looks and scores. Clipping a cow to highlight the positive qualities, such as a straight topline, is a strategy used by most producers.

The exercise in this chapter is designed to develop skill using the Dairy Cow Unified Scorecard to evaluate cattle.

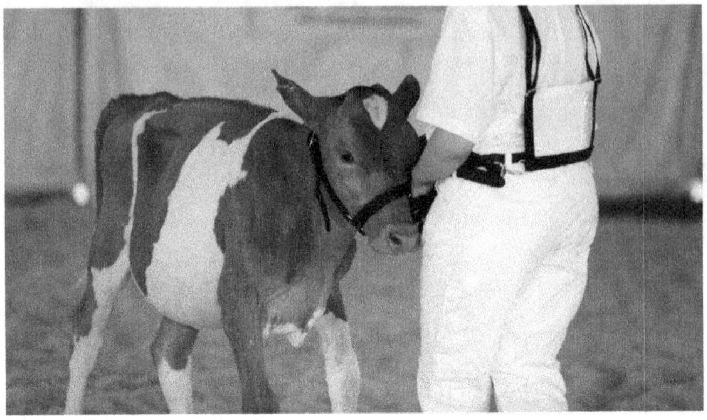

Youth dairy cattle show.

Name________________________________ Date__________________

Exercise 40-1 Dairy Cow Scorecard

When evaluating dairy for quality of conformation and breed traits, the Dairy Cow Unified Scorecard is used. In Part 1 of this exercise, identify the general desired characteristics that should be observed. In Part 2, describe the desired characteristics of each breed based on the scorecard. Finally, in Part 3, evaluate a group of dairy cattle using the scorecard. Use the unified scorecard located in the textbook for reference.

Part 1

Directions: Identify the desired characteristics for each of the areas of evaluation.

1. General appearance:

____

____

2. Dairy character:

____

____

3. Body character:

____

____

4. Udder:

____

____

Part 2

Directions: Identify the desired characteristics of each of the following breeds and the factors to be evaluated.

1. Ayrshire:

____

____

2. Holstein:

____

____

Name____ Date____________________

3. Milking Shorthorn:

____

____

4. Brown Swiss:

____

____

5. Guernsey:

____

____

Part 3

Directions: The instructor will provide a situation to evaluate a group of lactating dairy cows. Score each cow based upon the Dairy Cow Unified Scorecard, explain each of your scores and place the cattle in order based on the score they receive.

Record evaluation results here:

Name______________________________ Date______________

Chapter 40 Matching Activity

Directions: Match the term to the corresponding definition or description.

Term	Definition
____ 1. pedigree	**a.** an udder that is still firm after milking
____ 2. linear classification	**b.** the record of the animal's ancestors
____ 3. meaty	**c.** a modification of type classification that utilizes a computer program to score dairy cattle on a number of individual traits
____ 4. milkfat	**d.** the liquid remaining after milk has been curdled during cheese making
____ 5. dairy strength	**e.** production capacity
____ 6. dairyness	**f.** (pregnant) heifers often carry some surplus fat that may make them appear to be coarse over the withers
____ 7. blind quarter	**g.** the fat portion of the milk
____ 8. dry cow	**h.** are formed by the addition of the additives to milk to make it coagulate
____ 9. springing	**i.** a cow that is not producing milk
____ 10. curds	**j.** a section of the udder that does not secrete milk
____ 11. whey	**k.** indicates the milking ability of the animal, along with strength that supports good milk production

Name____ Date____________________

Chapter 40 Lab Questions

1. Describe the term *dairy* and how it is judged on a dairy cow.

2. Describe the steps that should be taken when judging dairy cattle.

3. Make a list of 20 terms that may be used when giving your oral reasoning for your dairy cattle score placings.

Chapter 41

Feeding Dairy Cattle

Introduction

Feed costs are about 40 to 50 percent of the total cost of producing milk. A total mixed ration (TMR) has all ingredients blended into one ration and allows a dairy cow to reach her full potential for milk production. A good, high-quality roughage is the basis of a ration, and grain and protein supplements make up the other portion of the animal feed. Many dairy farmers group cattle according to similar sizes and stages of production. This allows the farmer to mix a ration for each group that targets the needed energy requirements. This is much more efficient than trying to meet individual cows' nutritional requirements.

Hay, green chop, silage, haylage, and pasture are commonly used for roughage in dairy rations. Alfalfa hay and corn silage are two of the most common roughages used because they are the most economically produced. Corn and oats are the major grains used in the dairy industry. Protein supplements are essential in a TMR; some common examples of protein supplements are soybean oil meal, linseed meal, and cottonseed meal.

Dairy producers may use other products in the TMR. The use of these products depends on the availability of the product, the experience the farmer has had in the past, the price of the product, and the effect it has on milk production. A variety of by-product and other processed feeds may be used, including alfalfa meal, beet pulp, brewer's grain, and urea as a source of nitrogen.

The inclusion of minerals in the ration is very important for milk production and milk quality. Calcium, phosphorus, salt, and minerals are important and needed when balancing dairy rations. A commercial vitamin mix is usually mixed into the TMR. Vitamins A, D, and E are important components of a dairy ration. The most limiting factor in milk production is usually a shortage of energy in the ration. The stages of lactation, the size of the cow, and the amount of milk being produced per day are all factors used to determine the nutritional requirements.

A dairy calf is taken from the cow soon after birth. Its navel is dipped in iodine, and it is fed colostrum from the mother. Colostrum is the first milk produced by the fresh cow. It contains antibodies that help protect the calf from disease.

Name____ Date____________________

Exercise 41-1 Pearson Square

The use of the Pearson square was explained in detail in the balancing rations chapter of the textbook. Use the Pearson square method for mixing ingredients in a TMR.

Directions: For each of the following feed scenarios, determine the protein values needed for a productive TMR. Show all work used to calculate the ration.

1. Create 1 ton of a 16 percent protein ration using corn (10.9 percent) and soybean meal (41 percent). How much of each will have to be mixed together?

____

____

____

____

2. Make 1 ton of a 16 percent protein ration using alfalfa hay (18.7 percent) and oats (13.3 percent). How much of each will have to be mixed together?

____

____

____

3. Increase the protein levels and make 1 ton of a 23 percent ration using oats (13.3 percent) and cottonseed meal (41 percent). How much of each will have to be used?

____

____

____

4. Make a ration that is 22 percent protein. Use corn (10.9 percent) and dehydrated skimmed milk (35.8 percent). How much of each will have to be used to make 1 ton of this ration?

____

____

____

5. Make 1 ton of a 20 percent protein ration. Use sorghum (12.4 percent) and soybean oil meal (41 percent). How much of each will have to be used in the ration?

____

____

____

6. Mix a ration for dry cows that is 10 percent protein using prairie hay (5.8 percent) and a commercial supplement (36 percent). How much prairie hay and how much supplement will have to be used to make 1 ton of this feed?

____

____

____

Name____________________ Date__________

Exercise 41-2 Total Mixed Ration Lab

In Part 1 of this exercise, you will determine total mixed ration values. In Part 2, you will practice using the fist test to determine the moisture content of a TMR.

Part 1

Directions: Using the description of each cow and the major ingredients available, list the ingredients and values for a total mixed ration.

Cow 1

Description: Holstein heifer; 16 months; 900 lb; alfalfa mid-bloom

Cow 2

Description: Guernsey; 21 months; 1100 lb; grass

Name____ Date__________________

Cow 3

Description: Jersey; 5 months; 300 lb; budding alfalfa

Part 2

Directions: Determine the moisture content of various feeds provided by the instructor. Use your hand to squeeze a ball of the feed to estimate moisture content. Use the following fist test chart as a reference to determine moisture percentage.

Fist test	Moisture
• Stream of water	< 25% Moisture
• 6–10 drops	25–28% Moisture
• 1–5 drops	29–32% Moisture
• Ball doesn't open	33–35% Moisture
• Ball opens very slowly	36–40% Moisture
• Ball opens slowly	41–45% Moisture
• Ball opens rapidly	> 45 % Moisture

Name______________________________ Date______________

1. Feed description: ______________________________
   Estimated moisture content: ______________________________
   ____

2. Feed description: ______________________________
   Estimated moisture content: ______________________________
   ____

3. Feed description: ______________________________
   Estimated moisture content: ______________________________
   ____

4. Feed description: ______________________________
   Estimated moisture content: ______________________________
   ____

5. Feed description: ______________________________
   Estimated moisture content: ______________________________
   ____

6. Feed description: ______________________________
   Estimated moisture content: ______________________________
   ____

7. Feed description: ______________________________
   Estimated moisture content: ______________________________
   ____

Name____ Date____________________

Chapter 41 Matching Activity

Directions: Match the term to the corresponding definition or description.

Term	Definition
____ 1. challenge feeding	**a.** controls the amount of feed each cow receives
____ 2. transponder	**b.** refers to the amount of fat an animal is carrying
____ 3. double cropping	**c.** practice of feeding higher levels of concentrate to challenge the cow to reach her maximum potential milk production
____ 4. body condition scoring	**d.** growing two crops on the same ground in the same year
____ 5. total mixed rations (TMR)	**e.** one million calories
____ 6. dry cow	**f.** feed that has all ingredients blended into one ration
____ 7. hierarchy	**g.** ranking according to dominance
____ 8. windrowing	**h.** a cow that has ceased to produce milk
____ 9. megacalorie	**i.** raking together cut plant material

Name______________________________ Date______________

Chapter 41 Lab Questions

1. What are the main goals of a dairy cow feeding program?

2. What are the common ingredients used in dairy cow feeding rations?

3. What is a TMR and what are its advantages?

4. What procedures are essential to keep dairy calves alive after they are born?

Chapter 42

Management of the Dairy Herd

Introduction

Dairy herd and production records are the key to successful management and profitable production of milk. Dairy herd improvement associations (DHIAs) provide many records that help the dairy farmer increase net profits. DHIA records may add value to cattle and give producers a reference for improving or maintaining herd production levels. Dairy production records are standardized to a 305-day, twice daily, mature equivalent basis for comparison purposes. Several computer programs exist to help a dairy farmer properly manage the herd according to performance records.

Many modern dairy systems use microchip tags on the cattle that are read by a computer sensor. When the chip is scanned, information from each day is stored and can be used to track the performance of dairy cattle more efficiently. As a dairy cow ages and production begins to slow down, herd records can track the decrease in production and allow a herd manager to decide when to cull the animal from the herd. A regular culling program should be followed. Culling dairy cows is an easier decision when the price of beef is high because the dairy farmer captures more of the value of the cull cow.

It is most desirable for a dairy farm to raise its own replacement heifers. Not buying replacements can reduce the risk of introducing a new sickness to the herd. Herds that grow based on their own herd reproduction are called closed herds. Economically important traits should be considered when breeding cows for replacements. A proper breeding program should be established using a sire and cow that have a high genetic value and transmitting ability. Artificial insemination is commonly used on dairy farms because of the control over breeding schedule, the conception rates, and the ability to choose high-quality sires with ease. Catching the standing heat period in cows is essential for the artificial breeding program. Heifers being bred for the first time should be scheduled to calve at 22 to 24 months of age. A cow should be re-bred 50 to 60 days after giving birth.

In this chapter you will complete a record-keeping activity and a cull cow selection lab to further your knowledge in dairy cattle management.

iStock.com/mgstudyo

Veterinarian evaluating herd data.

Name______________________________ Date__________________

Exercise 42-1 Dairy Herd Record-Keeping Activity

You will complete a farm business analysis based on sample farm records provided by the instructor. The class will split into groups, and each group will be assigned a type of farm record to analyze.

Directions: Study the sample farm records for the area of data assigned. (1) Take notes on the data, (2) prepare a written analysis of the status of the farm in this area, and (3) prepare written suggestions for improvement based on the analysis of records.

- Production records on individual cows and on the herd
- Feed records
- Breeding and calving records
- Health records
- Cow identification records
- Financial records (the entire farm and enterprise)
- Inventory

1. Notes on the data

____

____

____

____

2. Analysis of records

____

____

____

____

3. Suggestions for improvement

____

____

____

____

____

Name________________________________ Date________________

Exercise 42-2 Cull Cow Selection Lab

A dairy farmer is milking 3,500 head of cattle. With so many cattle, the management of adding replacement heifers and removing cull cows for the herd is a full-time chore. A new crop of 500 replacement heifers is ready to be brought into the herd, and the herd manager has identified 480 cows to be culled for various reasons.

Directions: Based on the average percentage of culling due to various reasons (Table 42-1 from the textbook), determine how many cows were culled due to each reason.

Reason	**Number of cows to be culled**
Low production	____________
Reproduction problems	____________
Mastitis	____________
Disease	____________
Teat or udder injury	____________
Poor udder conformation	____________
Accidents and injury	____________
Poor feet and legs	____________
Other poor conformation	____________

Name____ Date____________________

Chapter 42 Matching Activity

Directions: Match the term to the corresponding definition or description.

Term	Definition
____ 1. estimated relative producing ability (ERPA)	**a.** total of all income received
____ 2. sexed semen	**b.** income (profit) left after expenses have been deducted
____ 3. net income	**c.** average genetic value for a certain trait that an animal transmits to its offspring
____ 4. gross income	**d.** groups of dairy farmers in the local area that provide production testing services
____ 5. dairy herd improvement association (DHIA)	**e.** prediction of a cow's production compared to that of other cows in the herd
____ 6. grade cattle	**f.** percent of the business the farmer owns debt free
____ 7. budget	**g.** available money to pay operating costs
____ 8. equity	**h.** calving difficulties
____ 9. capital	**i.** cows that are not purebred
____ 10. cash flow	**j.** become pregnant
____ 11. dry cows	**k.** cows that are down and cannot get up
____ 12. gestation	**l.** cows that are not producing milk
____ 13. downer cows	**m.** money available to invest
____ 14. parturition	**n.** begin to give milk
____ 15. dystocia	**o.** the period that a cow is receptive to breeding
____ 16. estrus	**p.** length of pregnancy
____ 17. freshen	**q.** estimate of income and expenses for a period, typically one year
____ 18. predicted transmitting ability	**r.** giving birth
____ 19. placenta	**s.** afterbirth
____ 20. settle	**t.** assures that over 90 percent of the calves born through artificial insemination will be of the gender selected

Name________________________________ Date______________

Chapter 42 Lab Questions

1. What are the main goals of a dairy producer?

2. What are the major reasons to keep accurate dairy production records?

3. What are the advantages of raising dairy herd replacements?

4. Describe the process of caring for a newborn dairy calf.

Chapter 43

Milking Management

Introduction

One of the most important parts of the cow is the udder. It is responsible for milk production and profitability; therefore, understanding the functions of the udder is important for dairy farmers. The udder is divided into four quarters. Each quarter has a teat that provides an outlet for the milk to be released from the udder. The alveoli in the udder wall are the component that manufactures milk. Milk letdown is the reaction of a cow to release stored milk. Gentle washing of the udder stimulates milk letdown. A hormone called oxytocin is also used to induce milk letdown.

Cows are creatures of habit, so developing a milking routine that all milking employees follow is important for maintaining a high level of milk production per cow. Cows should always be comfortable. Stress on dairy cattle will cause them to let down less milk, affecting the farm's profitability. A cow should not be under- or overmilked; either can cause discomfort and susceptibility to infection.

To produce high-quality, top-grade milk, several factors must be controlled. Milk quality can be maintained by keeping cows and facilities clean, rapidly cooling the milk, and properly cleaning the milking equipment. If these practices are not completed up to standards, bacteria can contaminate the milk and lower its quality. Good management practices help reduce bacteria in the milk. The level of bacteria can be measured by the presence of somatic cells. The somatic cell count of a dairy herd is a good indication of milk quality, herd health, and sanitary practices.

Milking equipment can become lined with mineral and milk stone deposits. Always follow manufacturer's directions to clean and sanitize milking equipment. Several sanitation factors affect milk quality. If the milk has any off flavors or odors, its quality is reduced, as well as its profitability for the farmer. Recognizing off flavors and identifying the cause are important skills to reduce quality loss.

In this chapter, you will complete a milk quality investigation.

Name______________________________ Date_______________

Exercise 43-1 Milk Quality Investigation

In this exercise, the class should be split into teams to practice the National FFA Milk Quality and Products Career Development Event (CDE). You can find information about this event on the FFA website (www.ffa.org).

Directions: Follow the directions in the following sample problem.

SAMPLE PROBLEM FOR MILK QUALITY AND PRODUCTS CDE TEAM EVENT

Teams will be given a table showing results of monthly tests performed on a single dairy farm and will perform four tests to complete the table. They will use the completed table to decide whether actions should be taken for violations of regulations or industry standards.

Teams need to fill in the blanks in the following sample table, prepare a report with reasons for the evaluation, and then present the report to a team of peer evaluators.

Test results for dairy farm #335522

Test	Month				
	1	2	3	4	5
Bacteria Count × 10^3	20	120	45	350	____
Somatic Cell Count × 10^3	100	500	650	750	____
Temperature (°C)	34	36	35	50	37
Antibiotic Test (+/−)	+	−	−	−	____
Freezing Point (°C)	−0.522	−0.500	−0.523	−0.493	−0.529
Titratable Acidity (%)	0.18	0.16	−0.21	−0.55	____

Courtesy of Future Farmers of America

Name______________________________ Date__________________

Record your team's report here:

Name____ Date__________________

Chapter 43 Matching Activity

Directions: Match the term to the corresponding definition or description.

Term	Definition
____ 1. off flavors	**a.** includes descriptions such as rancid, salty, flat, unclean, barny, and oxidized
____ 2. sphincter muscle	**b.** in the udder; contains specialized cells that synthesize milk
____ 3. alveoli	**c.** at the end of the teat; controls the flow of milk
____ 4. gland cistern	**d.** at the end of the teat
____ 5. streak canal	**e.** hormone secreted in conditions of stress
____ 6. milk letdown	**f.** occurs when the cow responds by a conditioned reflex to sensory stimuli such as the washing of the udder
____ 7. oxytocin	**g.** holds about 16 ounces of milk
____ 8. adrenalin	**h.** hormone released by the pituitary gland
____ 9. milking interval	**i.** a cup used to detect mastitis
____ 10. strip cup	**j.** time between milkings

Name______________________________________ Date____________________

Chapter 43 Lab Questions

1. Explain how a cow's udder produces milk.

2. What are some key factors that reduce stress for cows during milking?

3. How is the quality of milk assessed?

4. What factors can reduce the quality of milk?

Feel good studio/Shutterstock.com

Chapter 44

Dairy Herd Health

Introduction

Dairy herd health is the most important concern for a dairy farmer. If cattle are affected by health problems or disease, the farm's profitability decreases. If a sickness goes untreated, death of the animal is often the result. Carefully monitoring the health of a dairy herd and properly diagnosing and treating sickness must be a priority on a dairy farm. A veterinarian should be involved when necessary. Many commercial dairy farms hire a herd health manager who focuses on monitoring and treating herd health issues.

A herd health plan should be developed and implemented to maintain a healthy, profitable dairy herd. An effective plan emphasizes prevention of problems through regular vaccinations, testing for common infections and diseases, and good management practices. An important part of any herd health plan is keeping accurate records on each individual animal in the herd. These records should include health history as well as treatment, milk production, and reproduction records. Individual animal records are very helpful in accurately diagnosing and treating health problems.

Dairy cattle are at risk of infection from several diseases, bacteria, and parasites. A herd health manager must understand the symptoms and treatments necessary to control the impacts of infection and reduce the amount of profit loss. Some common health complications include displaced abomasum, retained placenta, ketosis, metritis, and milk fever, as well as internal and external parasites. Mastitis, or an inflammation of the mammary gland, is the most serious disease that affects dairy cattle because it directly affects milk quality.

Careful management and proper treatment of all herd health problems is needed to keep losses low. Using medications properly and allowing for withdrawal times is an important aspect of herd health.

In this chapter, a herd health plan will be developed for a sample dairy herd.

Name____ Date____________________

Exercise 44-1 Develop a Herd Health Plan Lab

Directions: For each herd health management area in the following list, describe the practices that should be followed to diagnose, treat, and reduce the spread of disease and parasites.

Herd Health Plan

Farm name: ____

Herd health manager name: ____

1. List eight general herd health practices that help reduce sickness. All employees should follow these practices when handling dairy cattle.

____

____

____

2. Biosecurity measures on the farm

Measures Taken with Incoming Animals	Practices Followed on Farm
• Sourcing policy	
• Ascertaining disease status	
• Pre-movement/purchase information from seller	
• Avoid/minimize mixing of stock	
Nominated Isolation Facility/Area	
• Location of nominated isolation facility or area	
• Means of cleaning and disinfecting after use	
Other Measures	
• People	
• Buildings	
• Equipment	
• Vehicles	
• Fencing	

Name____ Date__________________

3. Infectious disease control

Disease	Prevention/Treatment (vaccines)/Control Measures/Timing
Bruccelosis	
IBR	
BVD	
Parainfluenza-3	

4. Parasite control

Parasite	Prevention/Treatment(vaccines)/Control Measures/Timing
Fluke	
Lungworm	
Intestinal worm	
Lice	
Fleas	
Ticks	

5. Mastitis

Condition	Measures Taken	Product Used	Withdrawal Period
Environmental mastitis			
Contagious mastitis			

Name______________________________ Date____________

6. Other disease and health management problems

Condition	Prevention/Treatment (vaccines)/Control Measures/Timing
Milk fever	
Ketosis	
Displaced abomasum	
Retained placenta	

Name______________________________ Date______________

Chapter 44 Matching Activity

Directions: Match the term to the corresponding definition or description.

Term	Definition
____ 1. leukocytes	**a.** leukocytes and body cells
____ 2. *Bos taurus*	**b.** inflammation of the mammary gland
____ 3. placenta	**c.** white blood cells that fight infection
____ 4. mastitis	**d.** afterbirth of a cow
____ 5. dairy quality assurance program	**e.** second part of the ruminant digestive system
____ 6. Veterinary Feed Directive (VFD)	**f.** infection in the uterus
____ 7. veterinarian client patient relationship (VCPR)	**g.** designed to help dairy farmers produce high-quality milk while meeting regulatory standards
____ 8. over the counter (OTC)	**h.** first stomach of a ruminant
____ 9. ketosis	**i.** organization that made most antibiotics for livestock available by prescription only
____ 10. metritis	**j.** retail purchase with no need for a prescription
____ 11. rumination	**k.** nutritional disorder in dairy cows in which their blood sugar drops to a low level
____ 12. rumen	**l.** chewing cud
____ 13. cud	**m.** veterinarians that may prescribe medications for livestock if they believe the owner will follow their instructions
____ 14. reticulum	**n.** forage is chewed thoroughly for better digestion before re-swallowing and then regurgitating

Name______________________________________ Date____________________

Chapter 44 Lab Questions

1. Why is dairy herd health one of the most important areas of farm management?

____
____
____
____

2. What are some important methods in controlling and preventing diseases and parasites in dairy cattle?

____
____
____
____

3. List the most common diseases of dairy cattle.

____
____
____
____

4. What is the proper treatment if a cow is infected with mastitis?

____
____
____
____

Feel good studio/Shutterstock.com

Chapter 45

Dairy Housing and Equipment

Introduction

Dairy farms are very high in capital expenses. Some of the areas of the business that require investment are animal housing, milking facilities, manure handling facilities, labor expenses, and feed production and handling. Well-planned facilities save labor and make the operation more efficient in many ways. When designing a facility, the producer should address issues such as the location of the facility, the size of the planned herd, the laws and regulations that apply to dairy farms, the source and amount of money available, the type of milk market available, the amount of labor available, the kind of housing system to be used, the type of milking system to be used, the feed handling system to be used, the manure handling system to be used, and any future expansions that might be desired. Each of these areas must be thought of during the planning process; the goal should be efficiency, economy, and convenience.

The two most common types of housing used for the milking herds in the United States are stall (stanchion) barns and freestall barns. Most dairy farms that are milking more than 80 cows use freestall barns due to their efficiency. Many small dairy farms use tie-stall or stanchion barns. Many times, these buildings already existed; it is unusual for a modern dairy farm to build a new stanchion barn. Farmers that use a stanchion barn have a more labor-intensive milking system. Pail milkers, suspension milkers, and pipeline milkers are used in stall barns. Milking parlors with pipeline milkers may also be used in stanchion barns.

Freestall dairy barns are usually paired with a milking parlor. The most common type of milking parlor is the herringbone. Other types are side-opening, rotary or carousel, and polygon. Many freestall barns are equipped with automated equipment. This mechanization and automation requires a large capital investment, but the reduction in labor and increase in efficiency is proven to pay off when managed correctly. The use of crowding gates, power gates, feed gates, stimulating wash sprays, and automatic detaching units are just some examples of the automation available for modern milking parlors.

Between the freestall barn and the milking parlor, most designs include a concrete holding area to funnel cows into the milking parlor and contain them while they wait to be milked. The crowd gate is usually located in the holding area and helps push the herd through the parlor. Attached to the parlor is where a milk house is often located. A milk house contains the equipment for filtering, cooling, and storing the milk. A milk house is inspected by the milk inspector and must pass sanitary regulations. Large bulk tanks are located inside, or just outside of, the milk house. The bulk tanks are refrigerated and store milk at the proper temperature. With all of the mechanization of modern dairy farming systems, the energy costs can be very high. Good management practices reduce the energy bill by eliminating waste and exploring alternative energy sources.

The milking system uses a good portion of the total energy requirements of a dairy farm. It has several systems that must work properly to retrieve milk, cool it, and send it to the bulk tank for storage. A milking system is made up of the milking unit, the pulsation system, the vacuum supply system, and the milk flow system.

The calves and the older heifers on a dairy farm may be housed in buildings that require less capital investment. Either system works well with good management. Housing must be clean, dry, well ventilated, and free from drafts. Older heifers may be grouped in pens or kept in freestall barns.

Manure handling on a modern dairy farm is highly mechanized. Most manure is in liquid form. It may be stored in a lagoon until being spread as fertilizer on crop fields. Some farms have begun using methane digesters as a way to utilize the energy from the waste products.

Vladimir Mulder/Shutterstock.com

Modern carousel milking parlor.

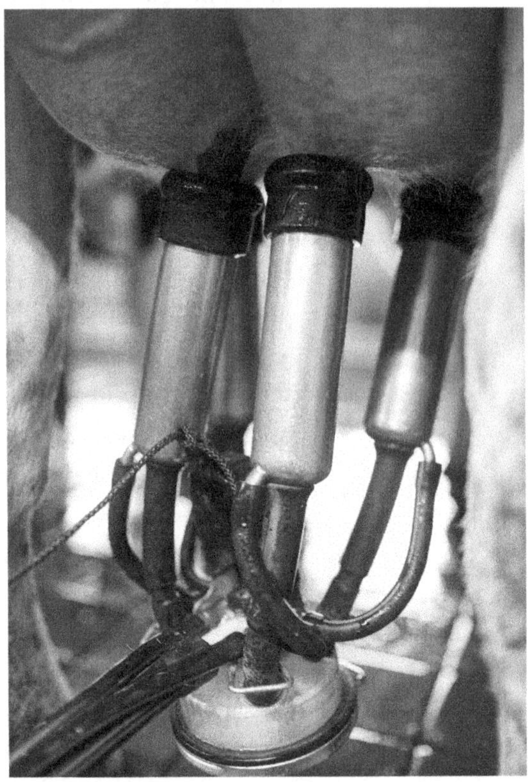

sergey kolesnikov/Shutterstock.com

Stanchion barn milking system.

Name______________________________ Date______________

Exercise 45-1 Design a Modern Cattle Handling Facility Lab

Two examples of dairy farms are provided here. Each description gives information about the size of the herd and systems that are used.

Directions: Read each dairy farm description. For each farm, draw an aerial view map of the facilities. Include each component of the farm in the drawing (that is, type of barn, milk house, parlor, manure handling, etc.). Be sure to label each building or structure. Use straight lines.

Example Dairy Farm 1

This farm has 1,100 head of cattle—600 are milking cows, 200 are calves, 200 are heifers, and 75 are dry cows, and there are 25 bulls on the farm that will be marketed for genetics. There are two barns used for the milking cows. Each barn has 275 stalls; there are fans located throughout the barn, misters to keep cows cool in the summer, large waterers throughout several areas of the barn, and an automatic scraping blade that slowly pushes manure into a gutter the end of the barn. The manure is then pumped from the gutter to a lagoon, where it is stored until there is time to spread the manure on the fields.

When it is milking time, the cows are moved from the freestall barn to the milking parlor and contained in a holding area while they are waiting to be milked. The milking parlor is a double-20 parallel parlor with automatic takeoff milking units. The milk is pumped to three separate 3,000-gallon milk tanks, where it is stored until the milk truck comes. The tanks are located just outside the milk house. The milk house contains a steam cleaner and sinks to keep equipment sanitary. The farm office is also located in the milk house. The calves are raised in calf hutches until they are three months old. From that time until they are lactating, they are housed in smaller confinement barns with a bedded pack. Each barn on the farm has a separate pen for sick animals that need treatment.

Example Dairy Farm 2

This is a small dairy farm with 110 head of cattle—75 are milking cows, and the other 35 animals are calves, heifers, and dry cows. A stanchion barn is used for the milking cows. The milking operator brings a milking unit to each cow and then attaches the unit to the vacuum and milk lines that run the length of the barn. There are two rows of cows facing away from each other with feed mangers in front of the cows and a manure gutter located behind them. Inside the manure gutter is a large chain run by a motor that moves the manure out of the barn and into a manure spreader that must be emptied onto the fields each day. The milk house is attached to the barn. It contains a sink to clean equipment and a 1,500-gallon milk tank in the center of the milk house.

Calves are in a small building attached to the back side of the stanchion barn. It has small box stalls for young calves or hutches and larger pens with a bedded pack for older heifers. There is also a box stall for sick cows.

Name____ Date__________________

Locate both design drawings in the space given below (include all needed facilities):

Name_______________________________________ Date________________

Chapter 45 Lab Questions

1. What major facility systems are needed on a dairy farm?

____

____

____

____

2. Which type of dairy operation is more efficient? Which is more economical?

____

____

____

____

3. List the various types of milking systems and parlors.

____

____

____

____

4. Describe the types of manure handling systems.

____

____

____

____

5. What type of housing is most used for raising calves?

____

____

____

____

Feel good studio/Shutterstock.com

Chapter 46

Marketing Milk

Introduction

The dairy market is constantly fluctuating throughout the year. Long-term trends have made it possible to predict when the price paid for milk will rise and drop. The long-term data show a seasonal trend in the dairy market. Dairy production levels complement milk price levels; the price of milk is usually higher when there is less production. The highest production occurs during the spring, and the lowest prices are paid in the six months following the spring.

Milk used for fluid product, cheese, butter, and ice cream makes up about 96 percent of the total milk processed every year in the United States. The consumption rate of dairy products, except for cheese, has shown a decreasing trend since the 1980s. Dairy advertising and education have been highly funded by promotional programs that aim to reverse this trend. If there is more milk on the market than demand, there are several protection programs in place to take product off the market in order to increase the price paid. An example of this strategy was the dairy herd buyout of the 1980s. This program took action to save a struggling dairy industry by paying farmers to sell their entire herds for slaughter. This took many cows out of production and lowered the supply of milk significantly. No actions this serious have been necessary since that time.

About 86 percent of the milk produced in the United States is marketed through cooperatives. Cooperatives provide many services to milk producers as well as to the entire market. Most of the cooperatives provide trucking of the milk from the farm to the processing plant. The cost of transportation has risen greatly over recent years due to rising fuel and labor costs. A fee for trucking is usually taken out of the farmer's milk check automatically.

Milk is divided into classes based on use for pricing purposes. Federal milk marketing orders are established in many areas to set prices for Grade A milk. The actual price paid is often above the minimum set in the federal order market.

Name______________________________ Date________________

Exercise 46-1 Dairy Advertising Activity

The dairy industry is known for its advertising campaigns. Dairy cooperatives and milk promotion agencies direct a lot of funding to producing effective advertising campaigns. The famous "Got Milk?" slogan is known by most Americans. A simple advertisement poster with a picture of a celebrity drinking a glass of milk and wearing a milk mustache has been very effective. In this activity, the class will be split into advertising teams. Each team will need to complete Parts 1 and 2. Prior to completing Parts 1 and 2, review several TV advertisements for milk to become familiar with common dairy commercials.

© s_bukley/Shutterstock.com

Part 1

Directions: Create an advertising poster that can be placed in grocery stores, in school cafeterias, on billboards, in magazines, on social media, and with other types of advertising outlets or sources. The poster should have some type of slogan to identify the objective of the advertisement and visuals that immediately make consumers think about milk products. Hand the poster in to the instructor for grading; then find a place in the school where the poster could be displayed. Posters should be completed on separate paper material and handed in to the instructor.

Name______________________________________ Date____________________

Part 2

Directions: The instructor has provided several examples of dairy products. Create a TV advertisement lasting approximately 30 seconds to promote the product chosen by the group. Record the advertisement to be shown to the class. Use the following form to prepare and plan for the advertisement.

Examples you can use: WeVideo, YouTube, TikTok, etc. Be creative!

Type of product	
Name of product	
Slogan	
Music	
Type of advertisement (funny, romantic, etc.)	
Script: conversation/voiceover	

Name____ Date____________________

Exercise 46-2 Milk Grading Lab

Background: Milk is classified into different types based upon the amount of fat. Skim milk contains 0 g of fat per 250-ml serving; 2-percent milk contains 2 g of fat per 250-ml serving; whole milk contains 8 g of fat per 250-ml serving; half and half cream contains 8 g of fat in a 2-Tbsp serving. Homogenization of milk breaks up the fat into very small fat globules and spreads them throughout the milk. The fat globules (0.1 to 15 μm in diameter) are basically suspended uniformly in the milk. Due to the fat content of milk, when it is mixed with a small amount of dish soap and food coloring, the milk begins to move and flow. The movement can be observed to determine which class of milk the sample belongs in.

Directions: Select four milk products from the following list. Create a hypothesis, follow the lab procedures, and record results in the data table; then conclude the lab based on the results.

Materials

Petri dishes	Liquid dish detergent	Toothpicks
Watch glasses	Assorted food colorings	Skim (fat-free) milk
1% milk	2% milk	Whole milk
Half and half	Buttermilk	Goat's milk
Powdered milk		

Hypothesis

Rank the milks by what you believe will result in the least to most movement after adding the food coloring. Provide an explanation (rationale) for your predictions.

Procedures

1. Place the four Petri dishes on the lab bench. Use a small piece of paper to label each dish.
2. Choose four of the eight milk choices provided for the lab.
3. Fill each Petri dish half full with the appropriate milk.
4. Equally space four drops of food coloring in each dish. Record your observations of the activity (movement) of the food coloring in the milk.
5. Dip a toothpick into the liquid dishwashing detergent. Then, touch and hold the toothpick into the middle of each dish.
6. Try again with more detergent, touching the milk in different areas. Record your observations of the activity (milk/food coloring movement) when soap is added.
7. Pour the used milk down the drain with lots of water. Clean each Petri dish with lots of soap and water; then dry with paper towels. Return all the materials and clean the lab area.

Name____________________________________ Date________________

Data Table

Type of Milk	Observations	Rank (least to most movement)

Conclusion

Write at least one paragraph describing the results of the lab and explaining the data. This should compare the hypothesis to what actually was observed during the experiment.

____

____

____

____

____

____

____

Name____ Date____________________

Chapter 46 Lab Questions

1. Describe the yearly trends for dairy production and pricing.

2. What has been the general trend for dairy consumption since the 1980s?

3. Describe some dairy promotion programs and how they promote increased consumption.

4. What is the difference between Grade A and Grade B milk?

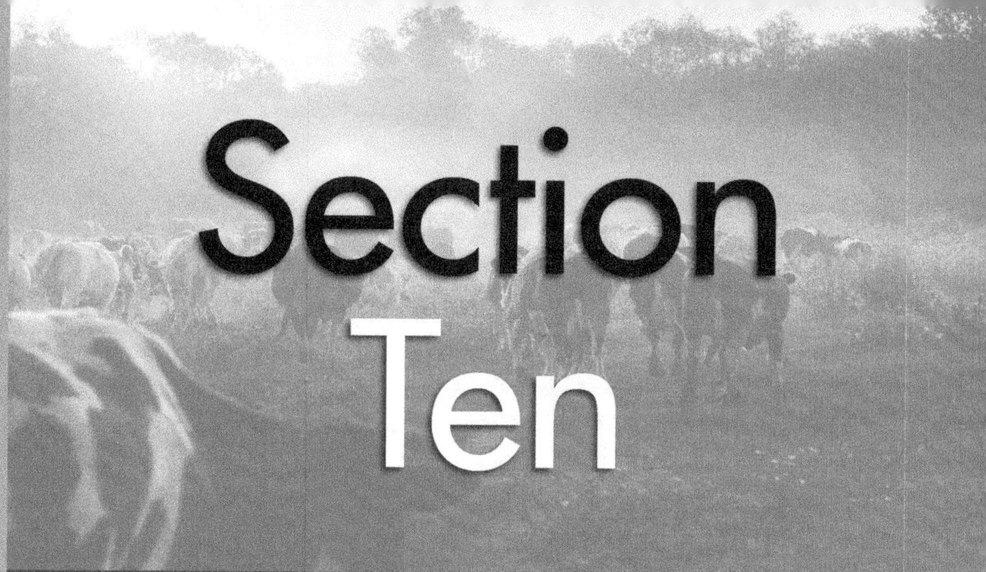

Section Ten

Alternative Animals

Feel good studio/Shutterstock.com

Feel good studio/Shutterstock.com

Chapter 47

Rabbits

Introduction

The rabbit is sometimes overlooked as a production animal. Rabbits are raised for meat, medical and educational research, pets, show, breeding stock, and for their fur or wool and skins. Modern breeds of rabbits have been developed and modified from European wild rabbits. The major use of domestic rabbits is for meat production. Rabbits produce a white meat that is high in protein and low in calories, fat, and cholesterol. The two major breeds used for meat production are the New Zealand and the Californian.

A feeding program is just as important in rabbit production as it is with any other type of livestock. Rabbits have a simple stomach, so plant products make up most of their diet. Most commercial rabbit producers as well as pet owners feed a completed pellet. There is not usually a need for additives or supplements in the pelleted ration. Just like other livestock, rabbits require fresh, clean water. Rabbits that are growing or used for reproduction have a higher energy requirement, so they are fed a completed pellet and have access to legume hay. Rabbits that are on a maintenance diet only require hay in their diet.

One of the advantages of the rabbit production industry is the short gestation periods of rabbits, which is 30 to 32 days. Female rabbits, called does, that are in good physical condition can be rebred 6 weeks after kindling. On an accelerated breeding program, a doe can produce up to five litters per year.

Rabbits are difficult to treat for disease; usually when symptoms are observed, treatment cannot save the animal. Good sanitation in the rabbitry is the key to disease control. Keep a clean environment and reduce the number of visitors to prevent disease transmission. Keep dogs, cats, birds, rodents, and insects out the rabbitry because they sometimes carry diseases or parasites that may be transmitted to the rabbits.

The exercises in this chapter reinforce an understanding of rabbit breeds and identification of rabbit health problems.

MestoSveta/Shutterstock.com

Californian production rabbit

Name______________________________ Date______________

Exercise 47-1 Breed Selection Activity

There are 50 breeds of rabbits according to the American Rabbit Breeders Association (ARBA) Standard of Perfection. Each breed has specific characteristics such as size, color patterns, physical attributes, and purposes.

Directions: For each of the following breeds of rabbit, research and record the breed characteristics and functions. In the square provided, paste a picture of the breed. These pictures may be cut out of magazines or found online and printed.

1. Checkered Giant
   a. Description ______________________________
   ____
   b. Function ______________________________
   ____
   c. Picture ______________________________
   ____

2. Dutch
   a. Description ______________________________
   ____
   b. Function ______________________________
   ____
   c. Picture ______________________________
   ____

Name________________________________ Date______________

3. New Zealand
   a. Description ______________________________
   ____
   b. Function ______________________________
   ____
   c. Picture ______________________________
   ____

4. Californian
   a. Description ______________________________
   ____
   b. Function ______________________________
   ____
   c. Picture ______________________________
   ____

5. English Lop
   a. Description ______________________________
   ____
   b. Function ______________________________
   ____
   c. Picture ______________________________
   ____

Name____ Date____________________

Exercise 47-2 Rabbit Health Problem Lab

There are many health problems that can affect rabbits. If quick diagnosis and treatment do not happen, death of the animal is often the result. In Part 1, students will practice matching symptoms with the health problems; in Part 2, students will perform physical examinations of rabbits looking for signs of health problems.

Part 1

Directions: Several examples of symptoms due to various rabbit illnesses are listed here. For each description of symptoms, (a) identify the health problem, (b) what caused the problem, and (c) the possible treatments.

1. Diarrhea containing blood. Slow to no weight gain or possible loss of weight. May be potbellied after recovery.
   a. ____________________________________
   b. ____________________________________
   c. ____________________________________
2. Loss of appetite; loss of weight. Diarrhea may be exhibited. Fur is rough. Pneumonia may develop.
   a. ____________________________________
   b. ____________________________________
   c. ____________________________________
3. Breasts become swollen and feverish; turn black and purple. Abscesses may form. Animal has fever.
   a. ____________________________________
   b. ____________________________________
   c. ____________________________________
4. Hair loss in circular patches; scaly skin with red raised crust. Matted fur. May occur on any part of the body.
   a. ____________________________________
   b. ____________________________________
   c. ____________________________________
5. Infected, inflamed, bruised, or abscessed bare areas on hind legs. Animal shifts weight to front legs. May affect front legs in severe cases. Secondary infections may occur.
   a. ____________________________________
   b. ____________________________________
   c. ____________________________________

Name________________________________ Date______________

6. External genitals and anus become inflamed. Crusts may form; bleeding may occur. Pus may develop in severe cases.

   a. ____________________

   b. ____________________

   c. ____________________

7. Dry, scaly, irritated skin; scratching; loss of fur on head, ears, and neck.

   a. ____________________

   b. ____________________

   c. ____________________

Part 2

Directions: The instructor will provide several sample rabbits. Give each rabbit a physical and visual examination. Palpate the body and pay attention to areas commonly affected by known health problems. If any health problems are detected, attempt to diagnose the problem, and then report the problem to the instructor.

Name____ Date____________________

Chapter 47 Matching Activity

Term	Definition
____ 1. herbivorous	**a.** refers to moving the hand gently back and forth, exerting a slight pressure between the hind legs and in front of the pelvis, to feel the fetuses
____ 2. coprophagy	**b.** false pregnancy
____ 3. kindling	**c.** ingestion of fecal matter
____ 4. pseudopregnancy	**d.** cage for keeping rabbits
____ 5. palpating	**e.** giving birth
____ 6. hutch	**f.** diet comes mainly from plant sources
____ 7. rabbitry	**g.** a mature or old domestic rabbit carcass of any weight, but usually over 4 pounds processed from a rabbit usually 8 months of age or older
____ 8. doe	**h.** male rabbit
____ 9. buck	**i.** having a single stomach
____ 10. kit	**j.** passing feedstuffs through the digestive tract a second time
____ 11. neutered	**k.** earth ware containers or metal troughs to feed rabbits in
____ 12. spayed	**l.** wild rabbits
____ 13. monogastric	**m.** place where rabbits are raised
____ 14. cecum	**n.** removal of the male rabbits' testicles
____ 15. enteritis	**o.** inflammation of the small intestine
____ 16. pseudo-rumination	**p.** a young rabbit suitable for harvest weighing 4.5 to 5.5 pounds
____ 17. crock	**q.** female rabbit
____ 18. fryer	**r.** female rabbit that has been fixed and unable to breed
____ 19. feral	**s.** baby rabbit
____ 20. roaster	**t.** a digestive organ in rabbits

Name____ Date____________________

Chapter 47 Lab Questions

1. Where did today's modern rabbit breeds descend from?

2. What breeds are the most popular for meat production?

3. What type of feed makes up a rabbit's diet?

4. How many litters can a rabbit have every year? How long is the gestation period of rabbits?

Feel good studio/Shutterstock.com

Chapter 48

Bison, Ratites, Llamas, Alpacas, and Elk

Introduction

The term alternative animals refers to animals and birds that are out of the ordinary, or not commonly used for livestock or poultry production. Alternative animals may include a wide variety of animals, but only the most common species will be covered in this chapter.

Bison are an unforgettable part of U.S. history. The American bison once numbered in the millions on the North American continent and were an important resource for the Indigenous people of the Great Plains. The settlers of the United States hunted the bison with no regulation or control, and the population was nearly brought to extinction by 1900. Wild and domestic populations of bison are now on the rise. A growing number of people are raising bison for sale as breeding stock and for meat. Bison production takes a great deal of knowledge and care as well as safe animal handling practices. Bison are very large, strong animals that are less tame than domestic breeds of cattle, so extra precautions must be taken when working with the animals. Bison need much less investment in housing and equipment than most domestic cattle, and they can grow on a similar feed ration.

Ratites are a group of flightless birds that includes the ostrich, emu, rhea, cassowary, and kiwi. The ratite industry in the United States has been growing, and the most common ratites for production are the ostrich and emu. These two species are the largest birds still in existence on Earth, and they are efficient converters of feed. The ostrich hen will usually produce 40 to 50 eggs per year, and the emu hen will produce about 30 eggs each year. Ratites are fed rations similar to poultry, but not enough research has been done to fully understand the nutrient requirements of ratites. One reason that more livestock producers do not raise ratites is that some of them can be very dangerous to handle and are aggressive toward humans. Without much research into ratite management, the best control of disease is prevention. There is not much medicine on the market specifically for ratites, and the population of veterinarians who will handle ratite cases is small.

Llamas and alpacas have a rich history in the Andes Mountains of South America, dating back 4,000 or 5,000 years. Llamas are used as pack animals as well as for meat, hides, and wool. Being smaller than llamas, alpacas are not usually used as a pack animal but are raised for their wool, which has softer fibers than either llama or sheep wool. Llamas and alpacas are modified ruminants, having three stomachs. The feeding program is based on the use of roughages, but the use of commercially prepared pelleted feeds is common. Llamas and alpacas are similar to horses in that they are very trainable and usually respond well to humans. Neither animal has an estrus period; rather, they can be induced into ovulation and bred during any time of the year. Herd health concerns are minimal for llamas and alpacas, but, as with any livestock animal, good sanitation and feeding programs will reduce health problems.

There are a growing number of American Elk farms across the country. Elk require very specialized handling practices and equipment. A producer must have experience and be very knowledgeable about elk behavior. Elk are primarily raised for their velvet, of which they may produce 40 to 50 pounds per year. Many Asian countries use the velvet for medicinal remedies. Elk are also bred for meat, by-products, and breeding stock.

In this chapter, students will compare management practices for bison to those used for domesticated livestock, calculate feed and water needs for bison, and identify major historical milestones for llamas and alpacas.

iStock.com/Gerald Zaffuts

Bison on fenced-in pasture.

Name____ Date____________________

Exercise 48-1 Bison Management Lab

Bison are very similar to cattle in their feed requirements and health concerns. Despite the similarities, bison are a less domesticated animal; they are larger, stronger, and faster than cattle. These traits make herd management more difficult—and potentially dangerous. Specific handling and management tasks should be planned according to the characteristics of bison.

Part 1

Directions: For each of the following areas of livestock management, identify three differences in management practices compared with other types of livestock.

Animal handling

1. ______________________________
2. ______________________________
3. ______________________________

Breeding

1. ______________________________
2. ______________________________
3. ______________________________

Weaning

1. ______________________________
2. ______________________________
3. ______________________________

Animal identification

1. ______________________________
2. ______________________________
3. ______________________________

Dehorning

1. ______________________________
2. ______________________________
3. ______________________________

Name____ Date____________________

Herd health

1. ____
2. ____
3. ____

Feeding practices

1. ____
2. ____
3. ____

Facilities and equipment

4. ____
5. ____
6. ____

Part 2

Directions: Read the following example of a feeding and water program for a small herd of bison. Answer the calculation questions using the information provided in the example.

Example: A bison will usually drink five gallons of water each day. In this herd, each bison eats three pounds of feed cubes, three times a week. This bison farmer feeds two 800-pound round bales to the herd each week. There are 68 head of bison in this herd.

1. How many round bales are needed for the entire year? ____________________
2. How many pounds of cubes does one bison eat in two weeks? ____________________
3. How much water does the entire herd drink in one day? ____________________
4. How many gallons of water does one bison need for a week? ____________________
5. How many pounds of cubes will be needed to feed the herd for one week? ____________________
6. A water trough holds about 80 gallons of water. If it fills up only once, how many water troughs will be needed to water the whole herd each day? ____________________

Name____ Date____________________

Exercise 48-2 History of Llamas and Alpacas Activity

Directions: List at least three major milestones for each period in the history of llamas and alpacas.

1. **Evolutionary history**

2. **South American history**

3. **United States history**

4. **Description of the modern llama and alpaca industry**

Name______________________________________ Date________________

Chapter 48 Matching Activity

Directions: Match the term to the corresponding definition or description.

Term	Definition
____ 1. blood typing	**a.** name for a baby llama, alpaca, or other camelid
____ 2. ratite	**b.** male parent of an animal
____ 3. cria	**c.** female parent of an animal
____ 4. sire	**d.** used to identify individual animals and to determine parentage
____ 5. dam	**e.** hairy, velvet-like skin on deer or elk antlers
____ 6. velvet	**f.** group of flightless birds that includes the ostrich, emu, rhea, cassowary, and kiwi
____ 7. alternative animals	**g.** no teeth on top
____ 8. niche market	**h.** refers to animals and birds that are out of the ordinary, or not commonly used for livestock or poultry production
____ 9. subtherapeutic	**i.** medications administered at lower-than-usual levels to improve productivity
____ 10. outbreeding	**j.** moving herds to different pastures during the grazing season
____ 11. rotational grazing	**k.** do not have estrus cycles
____ 12. monogamous	**l.** small part or subset of the larger market for meat
____ 13. polygamous	**m.** having one mate or partner
____ 14. camelids	**n.** members of the camel, llama, and alpaca family
____ 15. dental pad	**o.** having more than one mate
____ 16. induced ovulator	**p.** mating animals that are not closely related

Name______________________________ Date____________

Chapter 48 Lab Questions

1. What characteristics should one look for when selecting bison?

2. List four ratites and their uses.

3. What are different ways to group llamas and alpacas on a breeding farm?

4. Why is commercial deer farming steadily increasing?